# SOIL PROTOZOA

*Front cover photographs* (by W. Foissner) showing scanning electron micrographs of main types of soil protozoa. (a) *Mayorella* sp., a naked amoeba; (b) shell of *Nebela* (*Apodera*) *vas*, a Gondwanian testate amoeba; (c) *Bresslauides discoideus*, a colpodid ciliate; (d) *Polytomella* sp., a heterotrophic flagellate.

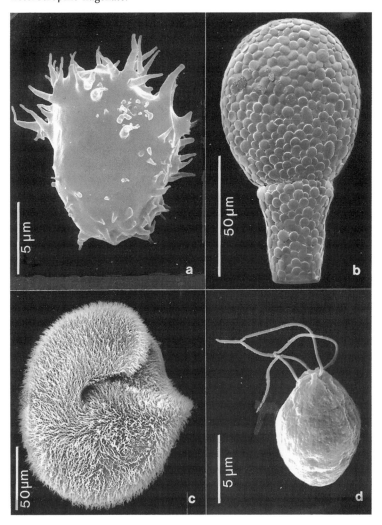

# SOIL PROTOZOA

Edited by

## J.F. DARBYSHIRE

*Honorary Associate*
*Macaulay Land Use Research Institute*
*and Honorary Research Associate*
*University of Aberdeen*
*Formerly Head of Department of Microbiology*
*Macaulay Institute for Soil Research*
*Aberdeen, UK*

CAB INTERNATIONAL

CAB INTERNATIONAL  Tel: Wallingford (0491) 832111
Wallingford        Telex: 847964 (COMAGG G)
Oxon OX10 8DE      Telecom Gold/Dialcom: 84: CAU001
UK                 Fax: (0491) 833508

A catalogue entry for this book is available from the British Library.

ISBN 0 85198 884 9

Phototypeset by Intype, London
Printed and bound in the UK by Biddles Ltd, Guildford

# Contents

v

Contents

# Contributors

Dr C. Alabouvette, *INRA Laboratoire de recherches sur la Flore pathogène et la Faune du sol, 17 rue Sully, BV 1540, 21034 Dijon Cedex, France.*

Dr A.J. Cowling, *Parthenon Publishing, Casterton Hall, Carnforth, Lancashire, LA6 2LA, UK.*

Dr J.F. Darbyshire, *Viewfield, Midmar, Inverurie, Aberdeenshire AB51 7QA, UK.*

Professor Dr W. Foissner, *Universität Salzburg, Institut für Zoologie, Hellbrunnerstrasse 34, A–5020 Salzburg, Austria.*

Dr B.S. Griffiths, *Soil–Plant Dynamics Group, Cellular and Environmental Physiology Department, Scottish Crop Research Institute, Invergowrie, Dundee DD2 5DA, UK.*

Professor T. Hattori, *Institute of Genetic Ecology, Tohoku University, Katahira, Aoba, Sendai 980, Japan.*

Dr P.J. Kuikman, *DLO-Research Institute for Agrobiology and Soil Fertility, Bornsesteeg 65, PO Box 14, 6700 AA Wageningen, The Netherlands.*

Dr P. Levrat, *INRA Laboratoire de recherches sur la Flore pathogène et la Faune du sol, 17 rue Sully, BV 1540, 21034 Dijon Cedex, France.*

Dr M. Pussard, *INRA Laboratoire de recherches sur la Flore pathogène et la Faune du sol, 17 rue Sully, BV 1540, 21034 Dijon Cedex, France.*

Dr J.A. Van Veen, *DLO-Research Institute for Plant Protection, Binnenhaven 12, PO Box 9060, 6700 GW Wageningen, The Netherlands.*

Dr K.B. Zwart, *DLO-Research Institute for Agrobiology and Soil Fertility, Oosterweg 92, PO Box 129, 9750 AC Haren, The Netherlands.*

# Foreword

This is the first book in English devoted entirely to soil protozoology for 65 years. It is a most timely addition.

The text of this book comprises a collection of essays written by experienced protozoologists. It is an unfortunate sign of the times that most of these authors will either have changed direction or retired by the time the current text is published. This loss of soil protozoologists does not seem logical when the US Department of Energy is investing many millions of dollars in microbial studies of the deep subsurface. Groundwaters, thought to be pristine until five years ago, have now been shown to be contaminated with waste waters, as indicated by the presence of protozoa. Similarly, studies on soil pollution and reclamation are being encouraged. Protozoa in the rhizosphere are being used to treat sewage in the reed-bed process. Furthermore, there is a trend in many countries to move away from the intensive use of mineral fertilizers and pesticides. Such changes in agricultural practice increase our dependence upon the soil organisms responsible for nutrient cycling and the biological control of pathogens and weeds. Interest in biodiversity is high in the minds of world governments, yet it is already difficult to find the expertise to identify soil protozoa and obtain advice on their ecology. Protozoa and nematodes are believed to be the major microfaunal secondary decomposers in soil and increased knowledge about their systematics and ecology is vital to food production and human well-being.

I am sure that future protozoologists will benefit greatly from these authors' efforts.

C.R. CURDS
*Keeper of Zoology*
*The Natural History Museum*
*London*
*UK*

# Preface

This book is a collection of essays about soil protozoa by a group of colleagues and friends. Several authors have had to overcome the difficulties of expressing their ideas in an alien language. I have asked the contributors to interpret their subjects in their own way, to identify deficiencies in our knowledge and to suggest promising topics for future investigation in the hope that this will stimulate enquiry among readers. As expected, the authors are not unanimous in their opinions. This book is not intended as a textbook, but it may serve as a modern introduction to soil protozoology for some soil microbiologists. No book in English has been devoted entirely to soil protozoa since the publication of Harold Sandon's classic *The Composition and Distribution of the Protozoan Fauna of the Soil* in 1927. It is hoped that a few readers will be encouraged to investigate these beautiful microorganisms for themselves and help to elevate soil protozoology from its present Cinderella-like status. Cinderella, the fairy tale, had a happy ending. Perhaps a similar fate awaits soil protozoology.

JOHN DARBYSHIRE

# Introduction 　1

J.F. DARBYSHIRE
*Viewfield, Midmar, Inverurie, Aberdeenshire AB51 7QA, UK.*

Protozoa were often detected in soils during the 19th century, but they were not then regarded as active components of the soil microfauna. It was only in the first decade of this century that a change of attitude occurred. Russell and his staff at Rothamsted Experimental Station in England provided some of the impetus for this change with their experiments on partial sterilization of soil. They discovered that plant yields increased more after attempts to sterilize soil samples by heating to 95°C rather than to 120°C, the normal temperature used for sterilization (Russell, 1917). Russell and Hutchinson (1913) interpreted this improvement in plant yields as due to the removal of some unfavourable biological factor, namely protozoa. They suggested that protozoa played a negative role in soil by devouring bacteria beneficial to soil fertility. Soon afterwards, active protozoa were extracted from soil for the first time by Martin and Lewin (1914, 1915) and the stage was set for increased interest in soil protozoa. Cutler *et al.* (1922) investigated the annual population fluctuations of six species of protozoa and bacteria in field soil using a long-term agronomic experiment at Rothamsted. They attempted to demonstrate relationships between protozoa and bacteria as well as protozoa and soil temperature or moisture. Although they found inverse relationships between the numbers of soil bacteria cultured on asparagine/mannite agar and the numbers of active amoebae determined by a most probable number (MPN) method in 86% of their daily observations, they failed to find significant correlations between these fluctuations and environmental conditions in the soil. Other investigations of mixed populations of microorganisms in soil have also shown that the situation in soil is far more complex than envisaged by Russell's simple hypothesis of good bacteria and bad protozoa (Singh and Crump, 1953). When this realization became widespread, interest in soil protozoan ecology waned and many protozoologists turned their attention to lethal parasitic protozoa.

Although Cutler and his colleagues were advised by a young statistican (R. A. Fisher), who was later to achieve international renown, they were handicapped by the computational facilities available to them in 1920. They recognized that further statistical analysis of their data might prove rewarding and accordingly included all their daily counts for one year in an appendix to their paper published in 1922. Recently, their data relating to *Cercomonas* spp. flagellates and bacteria have been reassessed after transformation into logarithms and using the cross-validatory quadratic smoothing technique of Stone (1974). This reanalysis has revealed previously undetected population oscillations that appear to be consistent with the hypothesis that total *Cercomonas* spp. populations estimated by MPN method were correlated with populations of bacteria cultured on asparagine/mannite agar within the confines of a range of soil moisture tensions and temperatures (Darbyshire *et al.*, 1993). *Cercomonas* spp. showed little growth below 5°C (Darbyshire *et al.*, 1993) and the flagellate ceases to grow at soil water tensions less than between –300 and –1000 kPa (Zwart and Brussaard, 1991). If the total populations of *Cercomonas* spp. from Cutler *et al.* (1922) data were plotted against their bacterial populations, Lotka-Volterra type population fluctuations were produced for the period before the soil began to dry out substantially (Day 310). Gause (1934) presented similar figures for predator/prey fluctuations for the ciliate *Paramecium aurelia* and the yeast *Schizosaccharomyces pombe* growing in laboratory microcosms. In Gause's figures, the plotted line usually moved gradually to what he referred to as the singular point near the centre of the circle. The cystic population estimates of *Cercomonas* spp. published by Cutler *et al.* (1922) were re-examined, but did not indicate any obvious trends. Recent investigations have suggested, however, that the HCl treatment method used by Cutler *et al.* (1922) does not reliably distinguish between cystic and trophic populations of protozoa (Sneddon, 1983; Pussard and Delay, 1985). A similar plot of Cutler *et al.* (1922) data for total counts of the amoeba *Naegleria gruberi* and bacteria is shown in Fig. 1.1. After Day 310 when the soil moisture fell below 18%, the plotted line moves towards the bottom right hand corner rather than towards the centre of the circle bounded by the plotted line between Days 190, 220, 250, 280 and 310. The hypothesis suggested earlier needs to be rigorously tested by monitoring microbial populations, moisture tension and temperature in the same soil horizon under field conditions to determine what proportion of the fluctuations in soil protozoan populations can be explained satisfactorily in this manner. Estimates of the edible bacterial population and modern alternatives to MPN methods of counting microorganisms may improve the correlation coefficients.

Interactions between soil protozoa and microflora are not only concerned with predator/prey relationships. For example, protozoan

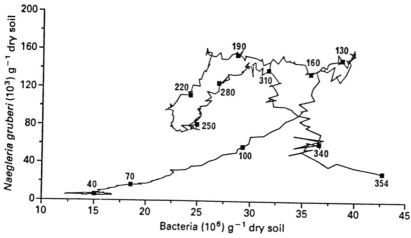

**Fig. 1.1.** The total populations of *Naegleria gruberi* and bacteria in Barnfield long-term experiment (1920–1921), Rothamsted Experimental Station (Cutler *et al.*, 1922) plotted against each other after cross-validatory quadratic smoothing. Time in days from 5 July 1920 is represented by the numbers on the figure.

metabolites are believed to stimulate bacterial metabolism (Nikoljuk, 1969; Huang *et al.*, 1981) and in return some fungal metabolites can lyse protozoa (Ebringer *et al.*, 1964; Darbyshire *et al.*, 1992). Detailed discussion of such interactions occurs in Chapter 6 by Drs Pussard, Alabouvette and Levrat. Some of our difficulties in assessing the significance of protozoa in soil are due to our ignorance about the environment in soil pores. Professor Hattori relates what is known about this environment to protozoan ecology in Chapter 3. Dr Cowling discusses the distribution of protozoa in soil profiles, habitats and continents in Chapter 2. His discussions emphasize the imperfections of protozoan methods and taxonomy. In the absence of any knowledge of the soil microbial environment, it is hardly surprising that Russell and his colleagues at first failed to realize the role of protozoa in nutrient cycling in soils. In Chapter 4, Dr Griffiths deals with the importance of protozoa as one of the major microbial groups of secondary decomposers involved in nutrient cycling in soils. Drs Zwart, Kuikman and Van Veen consider the same subject in relation to the small zone of soil close to plant roots, the rhizosphere, in Chapter 5. The use of protozoa as indicators of ecological change and soil quality are reviewed by Professor Foissner in Chapter 7. He has also suggested improved indices of protozoan diversity, a subject with important implications for the current debates of biodiversity. Lastly, I have attempted to foresee the future prospects for soil protozoology in Chapter 8.

# References

Cutler, D.W., Crump, L.M. and Sandon, H. (1922) A quantitative investigation of the bacterial and protozoan population of the soil, with an account of the protozoan fauna. *Philosophical Transactions of the Royal Society of London, Series B* 211, 317–350.

Darbyshire, J.F., Elston, D.A., Simpson, A.E.F., Robertson, M.D. and Seaton, A. (1992) Motility of a common soil flagellate *Cercomonas* sp. in the presence of aqueous infusions of fungal spores. *Soil Biology and Biochemistry* 24, 837–841.

Darbyshire, J.F., Zwart, K.B. and Elston, D.A. (1993) Growth and nitrogenous excretion of a common soil flagellate, *Cercomonas* sp. *Soil Biology and Biochemistry* 25, 1583–1589.

Ebringer, L., Balan, J. and Nemec, P. (1964) Incidence of antiprotozoal substances in Aspergillaceae. *Journal of Protozoology*, 11, 153–156.

Gause, G.F. (1934) *The Struggle for Existence.* Williams and Wilkins, Baltimore.

Huang, T.C., Chang, M-C. and Alexander, M. (1981) Effect of protozoa on bacterial degradation of an aromatic compound. *Applied and Environmental Microbiology* 41, 229–232.

Martin, C.H. and Lewin, K.R. (1914) Some notes on soil protozoa. *Philosophical Transactions of the Royal Society of London, Series B* 205, 77–94.

Martin, C.H. and Lewin, K.R. (1915) Notes on some methods for the examination of soil protozoa. *Journal of Agricultural Science (Cambridge)* 7, 106–119.

Nikoljuk, V.F. (1969) Some aspects of the study of soil protozoa. *Acta Protozoologica* 7, 99–109.

Pussard, M. and Delay, F. (1985) Dynamique de population d'amibes libres endogées (Amoebida, Protozoa) I. Évaluation du degré d'activité en microcosme et observations préliminaires sur la dynamique de population de quelques espèces. *Protistologica* 21, 5–15.

Russell, E.J. (1917) The recent work at Rothamsted on the partial sterilisation of soil. International Institute of Agriculture, Bureau of Agricultural Intelligence and Plant Diseases, Rome.

Russell, E.J. and Hutchinson, E.B. (1913) The effect of partial sterilisation of soil on the production of plant food. II. The limitation of bacterial numbers in normal soils and its consequences. *Journal of Agricultural Science (Cambridge)* 5, 152–221.

Singh, B.N. and Crump, L.M. (1953) The effect of partial sterilization by steam and formalin on the numbers of amoebae in field soil. *Journal of General Microbiology* 8, 421–426.

Sneddon, E.M. (1983) The interaction of protozoa and bacteria in a soil column. MSc thesis, University of Aberdeen.

Stone, M. (1974) Cross-validatory choice and assessment of statistical predictions. *Journal of the Royal Statistical Society* A 36, 111–147.

Zwart, K.B. and Brussaard, L. (1991) Soil fauna and cereal crops In: Firbank, L.G., Carter, N., Darbyshire, J.F. and Potts, G.R. (eds), *The Ecology of Temperate Cereal Fields.* Blackwell Scientific Publications, Oxford, pp. 139–168.

# Protozoan Distribution and Adaptation

**2**

A.J. COWLING
*Parthenon Publishing, Casterton Hall, Carnforth,
Lancashire LA6 2LA, UK.*

## Introduction

It might be expected that as protozoa are successful colonizers of many ecological niches in the biosphere, they would be widely distributed in most soils. Although it is rare to find a soil without any protozoa, terrestrial environments appear to impose exacting requirements on protozoa that invariably limit either their activity or ability to reproduce. These restrictions are usually associated with fluctuations in the supply of food or moisture.

Two distinct aspects of protozoan distribution in soil will be considered: firstly, the distribution and characteristics of protozoan *species* and *communities*; secondly, the distribution and abundance of *individuals* within species. Both aspects of distribution should be considered when investigating the underlying factors influencing protozoan distribution in soils. In this chapter, the scales of distribution are classified as either microspatial (μm to cm range), macrospatial (cm to m range) or zonal (km and above, i.e. the geographical range).

The frequency of occurrence of different species, most commonly in the form of data on presence or absence, has been collected from a variety of habitats. These data can be used to compare the *total* number of species and *species composition* in different sites. By determining *dominant, subdominant, common, infrequent* and *rare* species, it is possible to compare soil and non-soil faunas in terms of species diversity or 'richness'. Such comparisons help to suggest what ecological and physiological factors determine the establishment of any taxonomic group. The *frequency distribution* of species may indicate associations between habitat type and protozoan

5

morphological or physiological characteristics. Consistent distribution patterns of species amongst a range of soil types can form the basis for a classification of soil protozoan communities. Most work using this type of analysis has been done with Testacea (Bonnet, 1961, 1964a; Chardez, 1968). Bonnet, after extensive surveys of testate amoebae, adapted phytosociological categories of classes, orders, alliances and associations to testate communities (Table 2.1). Bonnet (1961, 1964a) distinguished the following testate groups in mineral soils: ubiquitous species (e.g. *Corythion dubium* and *Trinema lineare*), species characteristic of skeletal (immature or azonal) soils (e.g. *Assulina muscorum, Centropyxis aerophila* and *C. vandeli*) and species characteristic of mature zonal soils (e.g. *Centropyxis sylvatica, Phryganella acropodia, Plagiopyxis declivis, P. penardi* and *Trinema complanatum*). Bonnet used the term 'zonal' in the sense defined earlier in this chapter.

Plausible generalizations can be made by associating certain types of protozoan species and communities with edaphic characteristics, e.g. soil type, soil moisture regime and soil pore distribution. These characteristics often determine the nature of the microfloral component of the soil (bacteria, fungi, actinomycetes) that in turn provides the prey for the heterotrophic protozoan community.

Perturbations caused by human influence may be highlighted by so-called *bioindicator* species (see Chapter 7) and changes in species' distribution patterns may reflect significant differences between normal and perturbed habitats.

It is generally assumed that recording a species as present in a habitat implies its establishment in reasonably large numbers and the actual population size is often influenced by the same factors that affect its distribution. Thus, the *abundance* of individuals may reflect the suitability of the given range of these environmental factors for the species. The changing nature of many terrestrial habitats should not be overlooked; seasonal climatic factors are often of prime importance in determining, either directly or indirectly, the overall activity of soil protozoan populations (Bamforth, 1980). It is possible that even subtle differences in climate can be the principal factor preventing the establishment of a protozoan species in a particular region, e.g. the unexpected absence in the maritime Antarctic of the ubiquitous ciliate genus *Colpoda*. Although this genus could otherwise survive the annual 6–9 months of the Antarctic winter, the low *summer* temperatures appear to prevent the establishment of *Colpoda* populations (Smith, 1973, 1978).

**Table 2.1** Classification of soil testate rhizopod communities (from Stout and Heal, 1967, after Bonnet, 1961).

| Class | Order | Alliance | | Association |
|---|---|---|---|---|
| Soils in general *Phryganelletea* | Azonal soils low in organic matter *Centropyxidetalia* | Acid soils *Corythion dubii* | Soils with saxicole vegetation and roots | *Centropyxidetum deflandrianae* |
| | | | Skeletal soils | *Centropyxidetum vandeli* |
| | | | Soils very low in organic matter | *Paraquadruleto-hyalosphenietum insectae* |
| | | | Soils with accumulated organic matter | *Centropyxidetum plagiostomae* |
| | | Calcareous soils *Bullinuarion gracilis* | Decalcified clays | *Pseudawerintzewietum calcicolae* |
| | | | White rendzinas | *Geopyxelletum sylvicolae* |
| | | | Skeletal soils | *Arcelletum arenariae* |
| | Evolved soils rich in organic matter *Plagiopyxidetalia callidae* | Brown forest soils *Plagiopyxidion callidae* | Mor | *Plagiopyxidetum callidae* |
| | | | Mull | *Plagiopyxidetum penardi* |
| | | Grassland soils *Tracheleuglyphion acollae* | Humic alpine soils | *Tracheleuglyphetum acollae* |
| | | | Calcareous grasslands | *Centropyxidetosum elongatae* / *Tracheleuglyphetum acollae* |
| | Saline soils | Saline soils | Saline gleys | *Centropyxidetum halophilae* |

# Practical Difficulties in Determining Protozoan Distribution in Soils

Any satisfactory investigation of soil protozoan distribution must depend on a careful examination of a representative range of soil samples and reliable identification of the species present. Although the long-established principles of soil sampling may be adopted to ensure the statistical validity of the results, the quantitative methods for determining protozoan populations present in soil samples have serious inherent limitations (see Chapters 7 and 8). All methods require prolonged periods of microscopy (Lee and Soldo, 1992). Identifications are sometimes still based, especially in the case of flagellates, on descriptions made without the use of electron microscopy or phase contrast light microscopy. Microscopic examination of soil samples for living protozoa would appear, at first, to be preferable to more indirect methods. There are, however, some uncertainties with these so-called direct methods when applied to soil protozoa, because an unknown number of protozoa are either lost in the preparative stages, are obscured by soil particles, or are simply rendered unrecognizable in cases where there are few readily discernible morphological details for accurate identification. Recovery trials with laboratory cultures of protozoa added to soil do not necessarily reflect the behaviour of soil-grown protozoa intimately sorbed onto soil particles. In addition, direct methods involving living protozoa need to be completed before the protozoa have time to encyst or reproduce. This difficulty can be reduced by making many fixed and stained preparations, which can then be examined at leisure (Aescht and Foissner, 1992). Nevertheless, for testate amoebae, with their easily distinguishable and characteristic external shells, the direct method is generally preferred. The amoebal tests are concentrated by filtration (Coûteaux, 1967, 1975; Lousier and Parkinson, 1981) or flotation (Decloître, 1960; Grospietsch, 1965). Foissner (1987) recommends a compromise measure, also used by him for ciliates and flagellates, in which a thick (> 1 cm) layer of soil in a Petri dish is saturated with water. After application of a slight pressure, the excess water is sampled whilst tilting the dish.

The alternative *indirect* most probable number (MPN) method involves incubating dilution series of suspensions from soil samples for a prolonged period (4–6 weeks). The protozoa are allowed to multiply in this method and more time is available for identification (Singh, 1946; Darbyshire *et al.*, 1974). The total number of protozoa in the soil samples is calculated from presence or absence data of species in the dilutions. Formerly, encysted populations of protozoa were also calculated by treating duplicate soil samples with 2% HCl overnight before the dilutions were prepared (Cutler *et al.*, 1922). Recent investigations have suggested that this acid treatment does not reliably distinguish between cysts and trophozoites

(Sneddon, 1983; Pussard and Delay, 1985; Foissner, 1987). Heat treatment (45°C for 18.5 min) of the original soil suspensions has been used as an alternative method of destroying trophozoites (Rutherford and Juma, 1992), but comprehensive tests of this technique have yet to be published. The major disadvantages of the MPN method are that the protozoan cells may not be evenly dispersed in the diluent and that only a restricted range of protozoa may develop in the soil dilution series, rather than a representative range of all the protozoa present in the soils. This second disadvantage is particularly important when rich nutrient media are used. Casida (1989) clearly demonstrated that different MPN estimates of soil protozoa can be obtained, if different bacteria are supplied as food.

The practical difficulties of identifying species and estimating protozoan populations in soils are formidable barriers to progress in ecological studies. Comparisons of protozoan distribution in different soil samples are best confined to examples where the same quantitative methods are used throughout the comparisons.

## Mechanisms of Dispersal and Colonization

Soil protozoan distribution patterns at the micro- and macroscales reflect species and community responses to a large number of physiological and ecological factors. It should be emphasized that the patterns observed at any given time result from all the factors active *just prior to sampling.* Only repeated sampling will determine whether or not the same distribution occurs consistently. Each distribution is associated with a particular time-scale. At the microscale, changes may occur over periods of hours to days; at the macroscale, the process occurs over months to years, but global distribution has taken place over thousands to millions of years. In the first two scales the respective distribution patterns tend to be cyclic or irregular. With global distributions, the distances and time-scales involved are so large, relative to the size of the protozoa, that protozoan motility is insignificant and geographical barriers are the main factors restricting the spread of protozoa. Subsequent speciation is probably the main factor responsible for any differences between extant groups of related protozoa collected from different regions of the world.

Successful dispersal from an original inoculum, colonization of a new site and subsequent proliferation throughout this site by reproduction are the main stages by which protozoa are distributed amongst suitable soils. This occurs through protozoan motility or through the dissemination of viable propagules (cysts or trophozoites) by various agencies, such as air currents, water or other organisms, including humans. It has been estimated that there are an average of two protozoan cysts per cubic metre of air

(Sleigh, 1989). Wind dispersion may well be the means by which the most common soil species have spread throughout the world. Dispersion itself may be limited in some species either by an insufficient number of viable propagules or by other factors, such as propagule weight or geographical barriers. If so, the large-scale distribution patterns should reflect geographical and historical factors, such as remoteness from continental land masses and continental drift. The absence of a species may result from poor competitive ability (Maguire, 1963) and colonizing ability may reflect this. Recolonization by Testacea of surface litters after their destruction by deforestation or fire has been shown to be relatively slow, with some species requiring 6 months or more to re-establish themselves (Coûteaux, 1977; Buitkamp, 1979). Some species, notably *Euglypha rotunda* and *Corythion dubium*, appeared first. Significantly, *C. dubium* was also the only testate rhizopod to establish itself in relatively newly deposited (12-year-old) volcanic tephra on Deception Island (Smith, 1985) and the sole testate rhizopod to be recorded from sparsely vegetated fellfield soils in the maritime Antarctic (Smith, 1984). There have been few quantitative studies of colonization processes and most have concerned protozoan colonization of aquatic habitats (e.g. Cairns and Ruthven, 1972; Yongue and Cairns, 1978; Have, 1987). Small flagellates appear to be significant pioneer colonizers of soils, because of rapid en- and excystment, small size and rapid reproduction.

Compared with aquatic habitats, protozoan distribution in soils is more restricted. This is because of the heterogeneous nature of the soil matrix and the discontinuity of the water films surrounding soil particles. The scaling factor dictates that soil protozoa are more dependent on passive dispersal at the macroscale than at the microscale level. Bulk movement of protozoa down soil profiles may occur periodically after rainfall or irrigation, but subsurface lateral movement of protozoa relies in part on independent migration. Kuikman *et al.* (1990) consider such protozoan migration only occurs briefly when favourable moisture conditions prevail. In contrast, Rogerson and Berger (1981) noted rapid flagellate and amoebal colonization of sterile soil in buried perforated tubes, although this rapidity suggested that independent protozoan migration alone could not be responsible. Whilst the motility of soil amoebae is limited (*c.* 1–2 cm day$^{-1}$), the physical redistribution of surface litter and soil particles by raindrops, wind and other agents may greatly assist the colonization of soil surfaces (Lousier, 1982). Little is known of animal transport of protozoa within or between soils, although it must occur to varying degrees whenever soil is carried by migratory animals (Schlichting *et al.*, 1977; Chardez, 1986; Marciano-Cabral, 1988). Dissemination of soil protozoa may also occur via animal faeces. These protozoa survive ingestion and passage through the alimentary canal of the host animal either as cysts (e.g. *Bodo*) or trophozoites (*Copromonas*). This distribution pathway may have been significant in the evolution of coprozoic species into true endosymbionts, such as

*Trichomonas* spp. that are still able to survive outside the host (Stout and Heal, 1967; Bamforth, 1980; Foissner, 1991).

## Morphological and Physiological Characteristics of Soil Protozoa Affecting their Distribution

Several characteristics of soil protozoa that may be advantageous for life in soil are listed in Table 2.2.

### *Small size*

Small size may be regarded as a protozoan adaptation for rapid reproduction (Fenchel, 1974; Taylor and Shuter, 1981; Lüftenegger *et al.*, 1985; Wickham and Lynn, 1990) and exploration of the microscopic pores of soil. Comparisons between overall dimensions of individual ciliates from soil, freshwater and marine sands show that the soil isolates are generally the smallest (Foissner, 1987). The identical species from both soil and freshwater, however, do not show such differences. Foissner (1987) found no significant difference in mean body lengths and widths between soil and freshwater populations, although earlier workers had suggested that the reduced dimensions of soil forms resulted either from the relative nutrient deficiency

**Table 2.2.**  Morphological and physiological characteristics of soil protozoa.

Morphology:
    small dimensions
    dorsoventrally flattened and elongated
    test, if present, with small aperture to reduce protoplasmic exposure to
      desiccation
    body appendages:
      spines
      trailing flagella for close attachment to substratum
      reduced ciliature
    flexible body

Thigmotaxis and adhesion to substratum

Ability to utilize alternative diets

Chemotaxis

Effective and rapid ex- and encystment

Polymorphic life cycle

(Sandon, 1927) or from the reduced oxygen tensions in soil habitats (Adolph, 1929). The phenotypic expression of size is sometimes more highly variable in soils than in aquatic environments, as for example in some testacean genera, where distinct subpopulations distinguished by shell morphology occur in different habitats (Schönborn *et al.*, 1983; Lüftenegger *et al.*, 1988; Schönborn and Peschke, 1988, 1990; Lüftenegger and Foissner, 1991). In those testaceans that use soil material for shell construction, such as *Difflugia*, the nature of the shell is directly dependent on neighbouring material (Ogden, 1983). The shells of the same species from different habitats can accordingly vary in appearance and colour (Heal, 1963b; Schönborn and Peschke, 1990).

## Body shape

This characteristic has been particularly studied with the shells of testate amoebae. A morphological series can be arranged from more globular or disc-like tests in aquatic forms (e.g. *Arcella*) to more typically flattened, wedge-like shells in a range of soil species (Bonnet, 1961; Rauenbusch, 1987; Lüftenegger *et al.*, 1988). Bonnet (1975) has also described an associated trend of gradual reduction and concealment of the test aperture in a series of soil species. A weak correlation can be observed between the predominant shell form and type of environment, with soil moisture being the most critical limiting factor (Chardez and Lambert, 1981). Intraspecies variation has also been reported for testates, with specimens from drier habitats possessing smaller apertures (Heal, 1963b; Laminger, 1978; Schönborn, 1982). Morphologically distinct geographical races of testate amoebae, mentioned above, reflect important phylogenetic relationships within families (e.g. the Euglyphidae) and show also that colonization of a habitat series can occur by progressive adaptation (Schönborn and Peschke, 1988; Schönborn, 1989).

A further successful adaptation to the soil environment, particularly of naked amoebae and many soil flagellates, is evident in their ability to penetrate narrow pores and move along very thin water films (Bamforth, 1985, 1988; Foissner, 1991). This probably enables such protozoa to continue feeding in soils that have dried below field capacity.

## Motility

Protozoa are likely to exploit their food resource most effectively when there is a continuum of soil solution throughout the pore network. In wet soils, protozoa are likely to disperse through water films and colonize the soil pores more rapidly than under drier conditions. Such saturated conditions may be short-lived in many soils and it is clearly an advantage for protozoa to exploit these temporary conditions quickly. Besides harbouring microfloral prey, the soil solution may also constitute a nutritive medium in itself, especially near deposits of soil organic matter. Little is known, however, about the significance of osmotrophic nutrition of protozoa in field soils.

*Bodo saltans* (Fig 2.1c) is a common flagellate of both soil and freshwater habitats. It possesses two flagella and may be able to detect its food bacteria chemotactically (Brooker, 1971; Mitchell *et al.*, 1988). The longer posterior flagellum normally maintains contact with the substratum, where most of its prey is usually concentrated in many environments. The anterior flagellum is involved in guiding food bacteria into the buccal cavity and cytopharynx. *Bodo* can also become free-swimming and this is clearly an advantage in the rapidly changing environment of soil pores. The common soil flagellate *Spumella* (= *Monas*) also has motile and sessile forms (Belcher and Swale, 1976). *Spumella* can attach itself to the substrate by a posterior extension of the cell, rather than by a flagellum (Fig. 2.1d). Similarly, the cercomonad *Heteromita globosa* can temporarily attach itself to the substrate by means of a posterior filamentous extension of the body (Cowling and Smith, 1988). Protozoa that are sessile throughout their life cycle are comparatively rare in soils; even the peritrich ciliate *Vorticella*, which for most of the time remains attached to the substratum as a sessile zooid when feeding, has a motile telotroch stage.

### Adhesion and resistance to elution

The ability to resist displacement from beneficial sites in soil is a major characteristic of many soil protozoa. Most testate amoebae are not easily washed downwards and retain their original position in the soil, because their relatively large shells cannot pass through the constrictions of the soil pore network, e.g. the globular tests of *Centropyxis* and *Phryganella* spp. and the dorsoventrally compressed tests of *Assulina* and *Euglypha* spp. (Fig. 2.2). By far the most successful strategy, however, is strong adhesion to a wide variety of substrata. This ability enables the protozoan to remain in close contact with surfaces that are likely to be covered with thin water films and so conducive to prolonged protozoan activity. Adsorption on to soil particles, particularly clay, may be the key factor that enables smaller protozoa to remain *in situ* in the same way as bacteria, such as *Pseudomonas* spp. (Tan *et al.*, 1991).

The naked amoebae and mycetozoans have relatively large areas of contact with their substrate and adhere strongly to surfaces. The testate amoebae, however, are only in contact with the substratum near the shell aperture, but are still able to cling on tenaciously by means of pseudopodia that can often extend well beyond the length of the shell (Fig. 2.2). This

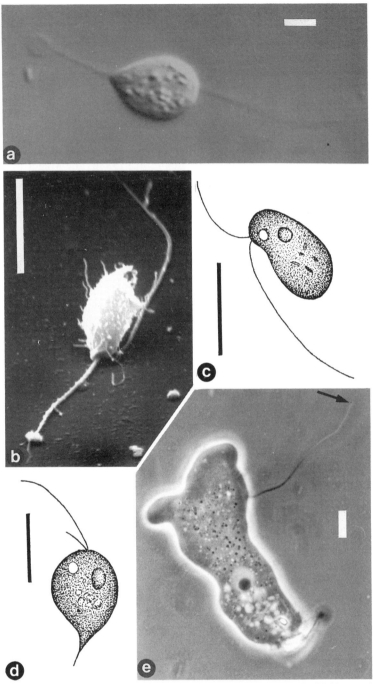

**Fig. 2.1.** Soil flagellates. (a) *Cercomonas agilis* (interference contrast); (b) *Heteromita globosa* (scanning electron micrograph); (c) *Bodo saltans*; (d) *Spumella (=Monas) elongata*; (e) *Mastigamoeba* sp. (phase contrast); arrow shows tip of flagellum. Scale bars represent 5 μm.

ability to adhere strongly to substrata may partly explain why both naked and testate amoebae are amongst the commonest forms of soil protozoa. Their filopodia and lobopodia are also able to penetrate soil pores too small for the whole organism to enter.

Most ciliates inhabiting marine sands have large, elongate and flexible bodies. Nevertheless, they are capable of rapid movement and are thigmotactic (Carey, 1991). Such attributes help these interstitial ciliates to resist elution and remain in the preferred surface layers of their habitat. Similar flexible vermiform shape and strong positive thigmotaxis amongst the hypotrichs, e.g. *Gonostomum, Histriculus, Holosticha* and *Oxytricha* spp., may enable these protozoa to resist displacement in a similar manner and explain why they are one of the more successful groups of soil ciliates. The spines of marine sand testaceans probably also provide some resistance against vertical displacement (Golemansky, 1977) and the spines on soil testacean shells may perform a similar function. Testate amoebae of moss-dominated soils, where the danger of displacement may be less serious, also show a tendency toward a loss or reduction of test spines (e.g. *Centropyxis*, Schönborn, 1989; *Euglypha strigosa*, Schönborn and Peschke, 1988).

## Protozoan diet

Sandon (1927) considered protozoan diet to be the principal ecological factor determining the overall composition and distribution of soil protozoa. It is to be expected that the distribution patterns of soil protozoa will be related to the local availability of their microbial food, but there is little information on this subject. Similarly, our knowledge of what constitutes the food of soil protozoa is rudimentary.

Protozoa show some specialization in their feeding habits and are selective in the food type they consume (Singh, 1942, 1945; Heal and Felton, 1970; Laminger and Bucher, 1984; Petz *et al.*, 1985). Bacterial feeding seems to be common, but a wider range of microorganisms is often ingested. For example, many soil ciliates feed on bacteria, but approximately half of the known species prey on other ciliates, flagellates or amoebae (Foissner, 1987). Many other studies have shown that protozoan growth is affected by the nature of their microbial diet (e.g. Cutler and Crump, 1927; Heal, 1963a; Taylor, 1979; Nisbet, 1984; Cowling, 1986; Mitchell *et al.*, 1988), and size-selective feeding by bacterivorous protozoa has been reported (Gonzalez *et al.*, 1990; Epstein and Shiaris, 1992). There is also some evidence for changes in protozoan food preferences in response to changing soil conditions and different soil types. The testate amoeba *Trinema enchelys* feeds mainly on organic detritus in drier soils, but when soil moisture is increased it ingests bacteria and other testate amoebae (Laminger, 1978). According to Bamforth (1985), protozoa and fungi show little association in soils, although antagonism between the two groups has been reported (Ebringer *et al.*, 1964; Darbyshire *et al.*, 1992). Also, increased numbers of the testacean *Phryganella acropodia* in soil microcosms enriched with fungi have been reported (Coûteaux and Devaux, 1983; Coûteaux, 1985). There are certainly some mycophagous species in most of the major free-living protozoan groups inhabiting soils (Chakraborty and Old, 1982; Petz *et al.*, 1985; Old and Chakraborty, 1986; Ogden and Pitta, 1990; Hekman *et al.*, 1992), but there are too few studies to assess the full ecological importance of protozoan-fungal interactions in soils (Levrat *et al.*, 1987, 1991) (see Chapters 5 and 6). Perhaps it is significant that mycophagous ciliate species have been reported mainly from soils in which fungi are abundant (Petz *et al.*, 1985).

Protozoa with specialized adaptations to feed on a narrow range of prey are rare in soils, although the ciliate genus *Grossglockneria* is a remarkable example (Foissner, 1980). This ciliate has a highly specialized feeding-tube that enables it to perforate fungal hyphae and spores (Aescht *et al.*, 1991). The ability of small flagellates to selectively graze bacteria attached to surfaces is another important adaptation to life in soils (Sibbald and Allbright, 1988). Schönborn and coworkers have observed that some testate amoebae engulf soil organic matter particles (Schönborn, 1965; Schönborn *et al.*, 1987).

## Encystation and quiescence

An important characteristic of many soil protozoa is the ability to survive adverse conditions either by encystment or transformation into a quiescent phase (Corliss and Esser, 1974). These physiological and morphological changes protect the protozoon from the more extreme fluctuations of moisture, temperature, pH, atmosphere and food supply in the soil environment. The main importance of encystment and quiescence for protozoan distribution is therefore to improve protozoan survival and to increase the possibility of wider dispersion. Those species that possess resistant cysts, combined with an ability for rapid excystment and encystment, tend to dominate soil protozoan populations, e.g. the common soil genera

---

**Fig. 2.2.** Testate amoebae of terrestrial habitats. (a) *Corythion dubium*, scanning electron micrograph, oblique view, showing shell constructed from small siliceous plates, with an invaginated aperture; (b) drawn from life, dorsal view, showing an active amoeba within its transparent shell, extending long, slender filopodia; (c) *Centropyxis aerophila*, scanning electron micrograph, showing globose test, invaginated and reduced aperture and agglutinate shell composed of cemented foreign particles; (d) *Phryganella acropodia*, side and ventral views showing agglutinate, opaque, hemispherical test, circular aperture and finger-like pseudopodia; (e) *Euglypha* sp., scanning electron micrograph of test showing dorsoventral flattening and pattern of overlapping shell plates. Scale bars represent 10 μm.

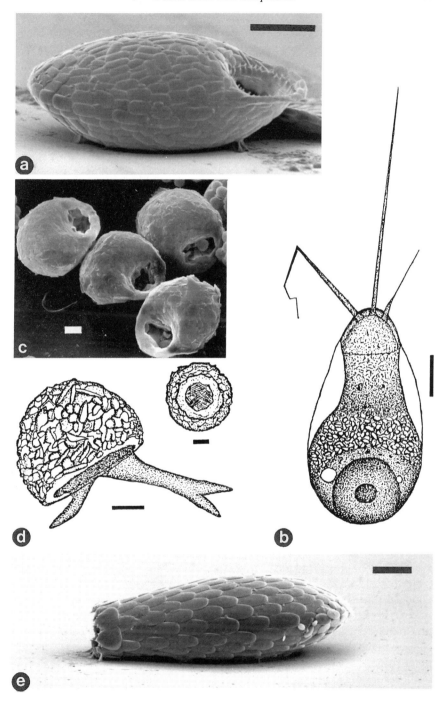

*Acanthamoeba, Cercomonas, Colpoda, Hartmanella, Heteromita, Naegleria* and *Trinema* (Stout and Heal, 1967; Schönborn, 1983; Laminger and Sturn, 1984; Bamforth, 1985; Smith *et al.*, 1990). The ubiquitous soil flagellate, *Heteromita globosa* (Fig. 2.1b), can encyst and excyst rapidly in response to changes in soil conditions. In a strain of *H. globosa* isolated from a maritime Antarctic soil, encystment and excystment were initiated by temperatures below and above a threshold of 1.5°C, respectively. Such soils have an average of 9 hours in the day when the temperature at 5 cm depth is above freezing even in summer. For survival, it is important for protozoa in this environment to excyst and feed rapidly when soil temperatures rise above freezing (Cowling and Smith, 1988; Hughes and Smith, 1989; Smith *et al.*, 1990). Other common soil flagellates, such as the chrysomonads *Spumella* (= *Monas*) and *Oikomonas* also readily encyst and excyst in soils.

A large reserve of protozoan cysts probably occurs in most soils and many cysts can survive in air-dry soil for decades (Corliss and Esser, 1974). *Naegleria* cysts can be transported by dust particles (Marciano-Cabral, 1988) and cysts of many other common soil amoebae have been recovered from the atmosphere (Rivera *et al.*, 1987).

After examining several cultivated soils, Foissner (1987) suggested that most ciliates are encysted under most field conditions, because he found large numbers of protozoa in air-dried, remoistened soil and few active protozoa in fresh samples using direct counting methods. This apparent suppression of activity in soils he termed 'ciliatostasis', by analogy with the term 'fungistasis' used for soil fungi. If such phenomena are widespread, they could have important effects on protozoan distribution in soils (Petz and Foissner, 1988). It also emphasizes the importance of distinguishing between the numbers of active and encysted protozoa in soils.

Evidence from studies of rhizopods suggests that naked amoebae encyst readily when soil moisture tensions fall below *c.* −0.15 MPa (Alabouvette *et al.*, 1981; Bamforth, 1985; Pussard and Delay, 1985), but that Testacea can remain active for longer periods at these moisture tensions (Stout and Heal, 1967). The narrow apertures of testacean shells help to reduce desiccation of the cytoplasm. Where drought or starvation persists, most testate amoebae can finally encyst. Some testate amoebae (e.g. *Bullinularia, Plagiopyxis, Tracheleuglypha*) can also secrete an internal protective membrane (epiphragm) and are able to survive several months without food in this temporary precystic form (Bonnet, 1964a; Coûteaux and Ogden, 1988). The naked rhizopod *Thecamoeba* achieves great resistance to desiccation by becoming dormant without encystment (Page, 1976). A true and efficient encysted stage, however, is the strategy adopted by the majority of soil protozoa.

The complete absence from soils of some ciliate genera commonly found in freshwater habitats (e.g. *Paramecium*) and the apparently restricted distribution of the cryptomonad and euglenoid flagellates in terrestrial habitats may both be attributed to an inability to form resistant cysts. Euglenoid flagellates do, however, have the ability to survive temporary adverse conditions in a non-cystic resting form (Leedale, 1985). The significance of euglenoids in soils may possibly have been underestimated (Foissner, 1987).

## Life cycle

A more advanced strategy for dispersion, from the usual situation of resistant cysts and active trophozoites, is one where further morphological differentiation occurs in the life cycle. *Naegleria gruberi*, a common soil amoeba, has three morphological forms (Fig. 2.3). With abundant moisture, *N. gruberi* can transform from an amoeba into a biflagellated cell (Marciano-Cabral, 1988; Page, 1988). Other amoebal genera that have flagellated stages include *Paratetramitus*, *Tetramitus* and *Vahlkampfia* (Page, 1988). In a few protozoa, both flagella and pseudopodia are present in the same cell, as in *Mastigamoeba* sp. (Fig. 2.1e).

Several other soil protistan genera have evolved polymorphic life cycles; the best known examples are the slime moulds (Eumycetozoa) (Feest, 1987; Blanton, 1990). For example, the cellular slime mould *Dictyostelium* exists as either a solitary amoeba, a cyst, a migratory 'slug' of many cells, a fruiting body, or a spore at different stages of the life cycle. Such a life cycle may be regarded as an elaborate dispersal mechanism. By such means, the microbial food supply is exploited over a wide area; the disadvantage appears to be greater reliance on a large number of environmental cues required for successful completion of the life cycle. *Pocheina flagellata* is the only cellular slime mould (Acrasidae) known to have flagellated cells. The syncytial slime moulds (Myxogastria), such as *Physarum*, have even more complex life

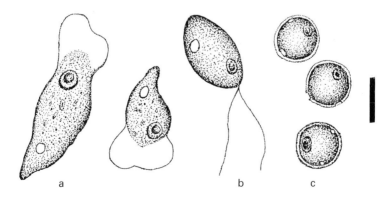

a            b     c

**Fig. 2.3.** The common naked soil amoeba *Naegleria gruberi*. (a) Trophozoites; (b) flagellate; (c) cysts. Scale bar represents 10 μm.

cycles, with flagellates, amoebae, plasmodia, sclerotia and spore-forming structures present at different stages (Sleigh, 1989). A remarkable example of polymorphism in the Ciliata is shown by the epiphytic hymenostome *Sorogena stoianovitchae*, which feeds exclusively on *Colpoda* spp. Its life cycle is an extraordinary case of parallel evolution to that of slime moulds (Olive and Blanton, 1980; Blanton and Olive, 1983). When its food supply is exhausted, *Sorogena* can transform from a free-swimming carnivorous form into an aerial fruiting body or sorocarp by the aggregation of many individual cells. When this fruiting body dehydrates, it splits, releasing spore-like cysts.

This increasingly complex series of life cycles may be regarded as a series of progressively increased adaptations to the terrestrial environment.

## Environmental Factors Determining Spatial Distribution Patterns

The major factors that affect the spatial distribution of soil protozoa within soils are listed in Table 2.3. Some (e.g. soil type or vegetation) often have a major influence on the protozoan community present; others (e.g. soil microclimate) may have a less profound influence. The combined influence of all these factors ultimately determines the success of any given species. It may often be difficult, however, to distinguish between the major factors determining soil protozoan distribution patterns or even whether a consistent pattern is evident over the complete range of each factor. It is possible to classify protozoan communities in terms of moisture regimes (Stout, 1974) as aerophilic (in areas prone to drought), hydrophilic (permanently aquatic habitats) and osmophilic (saline habitats), but such a classification only provides general guidelines.

**Table 2.3.**   Environmental factors determining protozoan distribution in soil.

| |
|---|
| Soil type |
| Geometry of soil pores |
| Organic matter content |
| Soil pH, temperature, moisture and atmosphere |
| Vegetation cover (above and below ground) |
| Litter and humus type derived from plants |
| Root exudates |
| Distribution of microbial food |

# Microspatial Distribution Patterns

The heterogeneous nature of the soil matrix at the microspatial level (μm to cm range) results in discontinuous distributions of protozoa. In many instances, the critical factor determining the persistence of protozoa in any location in soil will probably be access to microbial food, which is often associated with deposits of organic matter. Although these microsites are likely to be scattered throughout the soil, the general rate of input and vertical distribution of soil organic matter depends on the major form of fresh organic matter added to soil from plant, animal or faecal origins. The recalcitrance of this organic matter to decomposition will determine its residence time in the soil, but most of it will be rapidly exploited by the complex microbial and soil faunal communities that constitute the decomposer food webs. As the majority of soil protozoa are heterotrophic and particulate feeders, most will flourish at the expense of the bacteria near the sites of active decomposition.

The rhizosphere, considered in detail in Chapter 5, can be regarded as a model dynamic system in which most of the factors involved in determining overall soil protozoan distribution and activity at the microspatial level are at work, i.e. protozoan dimensions, chemotaxis, motility, food selection, soil-pore dimensions and interconnectivity, root exudates, soil moisture, nutrients and the nature of the soil microflora. The structure of soil aggregates and the nature of the soil-pore network largely determine the physical nature of the soil microenvironment of protozoa. The protozoan stage in the food web may be considered as an intermediate trophic level that exploits a range of soil pores and food to which other important soil invertebrates (i.e. nematodes, rotifers, microarthropods) have relatively less access. The importance of these metazoa, both as predators of protozoa and competitors for microbial food, should not be overlooked (Anderson *et al.*, 1978; Elliot *et al.*, 1980; Griffiths, 1990).

# Macrospatial Distribution Patterns

The principles governing macro- and microspatial distribution patterns are often the same, so similar distribution patterns may be evident at both scales. Thus, horizontal and vertical patterns at the macrospatial level would be expected to be aggregative, because they should reflect the heterogeneous nature of the soil-pore network and the limited availability of suitable food. A few studies have shown that protozoa have an aggregated distribution (Stout, 1962), but there is very little detailed information about between-site differences in horizontal macrospatial distribution patterns. Since the most striking distributions are likely to be in the vertical plane, most investigators have bulked soil-core samples together and selected discrete sample depths for comparison in order to obtain frequency data (Feest, 1987; Foissner, 1987).

Vertical patterns of protozoan distribution in soil result from the nature of the habitat, particularly the nature of the vegetation associated with the soil, the humus and soil type. Most of the fresh organic matter will be added close to the soil surface and its declining distribution with depth in the soil usually influences protozoan distribution. The soil moisture regime and the related microclimate are also influential factors. Unfortunately, as these factors are all closely interdependent, it is often difficult to distinguish the major one responsible for a particular protozoan distribution, as discussed earlier. There are probably as many different distributional patterns as there are habitat types, although some generalizations can be made. In general, the highest numbers of protozoa are usually recorded in the upper 10 cm of soil (Stout and Heal, 1967; Foissner, 1987), although estimates of numbers or biomass of protozoa in soil habitats vary considerably (Table 2.4).

### Mineral soils

Most attention has been paid to testacean distribution patterns for the reasons indicated above and especially in woodland soils (e.g. Coûteaux, 1972; Chardez and Lambert, 1981; Rauenbusch, 1987). Bonnet's classification of testacean communities (see Table 2.1) is a useful guide to the characteristic species present. The vertical distribution of Testacea in a typical temperate mineral soil under a mixed woodland shows a fundamental difference between the type of species inhabiting the L layer and the other holorganic layers at the soil surface (i.e. fermentation (F) and humus (H) layers: Berthelin and Toutain, 1982). The superficial L layer is subject to greater desiccation and water is usually restricted to a thin film, whereas more moisture is available in the F and H layers. The species in the L layer tend to be those with laterally compressed tests, whilst those in the F and H layers have more globular tests. In general, however, the holorganic soil horizons are inhabited by testate amoebae with large pyriform tests, such as *Heleopera* and *Nebela*, whilst the layers of mineral soil tend to contain species with smaller tests. Testacean genera frequently found in the upper mineral soil layers are *Centropyxis*, *Corythion*, *Euglypha*, *Plagiopyxis* and *Trinema*. Several other studies have shown a pronounced stratification of testacean species in forest soils (Volz, 1951; Bonnet, 1961; Schönborn, 1962; Stout, 1963; Chardez, 1964; Stout and Heal, 1967; Schönborn *et al.*, 1987) and this is apparent also for ciliates (Stout, 1963). Band (1984), however, detected no species differences amongst naked amoebae between different sites or soil horizons in northern hardwood forest soils in the USA,

**Table 2.4.** Published data for soil protozoan density and biomass from selected references, showing the range and variability of estimates from a variety of habitats. Figures are annual means unless otherwise stated (DW = dry weight; WW = wet weight of soil).

| Reference/soil type | Density | Biomass |
|---|---|---|
| **TESTACEA** | | |
| Foissner, 1987 | | |
| alpine mat (moder) | $20 \times 10^6$ m$^{-2}$ | 1.165 g m$^{-2}$ |
| alpine rendzina (moder) | $39.6 \times 10^6$ m$^{-2}$ | 2.209 g m$^{-2}$ |
| wheat fields/meadows | | |
| conventionally farmed | 751 g$^{-1}$ DW | 0.0279 mg g$^{-1}$ DW |
| organically farmed | 878 g$^{-1}$ DW | 0.0375 mg g$^{-1}$ DW |
| Heal, 1962 | | |
| uniform sward of *Sphagnum* | $16 \times 10^6$ m$^{-2}$ | *c.* 1 g m$^{-2}$ |
| Lousier, 1982 | | |
| aspen woodland soil (L,F, H, Ah layers) | 93,483 g$^{-1}$ DW | 0.4286 mg g$^{-1}$ DW |
| Lousier and Parkinson, 1984 | | |
| aspen woodland (mor) soil | $261 \times 10^6$ m$^{-2}$ | 0.723 g m$^{-2}$ |
| Meisterfeld, 1977 | | |
| *Sphagnum* fen/bog sites | $13.6 \times 10^6$ m$^{-2}$ | 0.7–4 g m$^{-2}$ [a] |
| Meisterfeld, 1986 | | |
| beech woodland (mull) soil | $84 \times 10^6$ m$^{-2}$ | 1.715 g m$^{-2}$ |
| Meisterfeld, 1989 | | |
| submontane beech forest/litter | – | 0.343 g DW m$^{-2}$ |
| Schönborn, 1986a | | |
| raw humus, spruce forest soil | $91 \times 10^6$ m$^{-2}$ | 1.922 g m$^{-2}$ |
| Smith, 1992 | | |
| continental Antarctic moss/lichen[b] | 50–2500 g$^{-1}$ WW | – |
| Smith and Tearle, 1985 | | |
| maritime Antarctic fellfields | 250–450 g$^{-1}$ DW | – |
| **CILIATES** | | |
| Bamforth, 1984 | | |
| arid desert soils | 110–460 g$^{-1}$ DW | – |
| arid desert litters | 340–9480 g$^{-1}$ DW | – |
| Bamforth, 1991 | | |
| beech woodland soil[c] | 571 g$^{-1}$ DW | – |
| ryegrass–clover pasture[c] | 35–167 g$^{-1}$ DW | – |
| Foissner, 1987 | | |
| alpine grassland soils | $0–4 \times 10^6$ m$^{-2}$ | – |

**Table 2.4.**  *Continued*

| Reference/soil type | Density | Biomass |
|---|---|---|
| wheat fields/meadows | | |
|   conventionally farmed | 1285 (0.2[c]) g$^{-1}$ DW | 27.5 mg g$^{-1}$ DW |
|   organically farmed | 1027 (1.4[c]) g$^{-1}$ DW | 19.8 mg g$^{-1}$ DW |
| **NAKED AMOEBAE** | | |
| Clarholm, 1981 | | |
|   podsolized pine forest soil[d] | 10$^5$–2 × 10$^6$ g$^{-1}$ DW | *c.* 2 mg g$^{-1}$ DW |
| Meisterfeld, 1989 | | |
|   submontane beech forest/litter | – | 1.133 g DW m$^{-2}$ |
| Napolitano, 1983 | | |
|   sandy beach soil | 79–585 g$^{-1}$ DW | – |
| **FLAGELLATES** | | |
| Bamforth, 1984 | | |
|   arid desert soils | 140–2320 g$^{-1}$ DW | – |
|   arid desert litters | 220–7930 g$^{-1}$ DW | – |
| Davis, 1980 | | |
|   Antarctic moss peats[e] | – | 1.5 mg DW m$^{-2}$ (0.8–2.5) |
| Foissner, 1991 | | |
|   spruce forest litter[c] | 7–30 × 10$^6$ m$^{-2}$<br>2–10 × 10$^3$ g$^{-1}$ DW | 3–12 mg DW m$^{-2}$ |
| Meisterfeld, 1989 | | |
|   submontane beech forest/litter | – | 0.054 g DW m$^{-2}$ |
|     upper (0–3 cm) layer[e] | 1.6 × 10$^9$ m$^{-2}$ (0.083–18.9) | – |
|     lower (3–6 cm) layer[e] | 1.2 × 10$^9$ m$^{-2}$ (0.061–7.3) | – |
| Persson *et al.*, 1980 | | |
|   Swedish pine forest soil | 40 × 10$^6$ m$^{-2}$ | 16 mg DW m$^{-2}$ |
| Smith and Tearle, 1985 | | |
|   maritime Antarctic fellfields | 80–3400 g$^{-1}$ DW | – |
| **TOTAL PROTOZOA** | | |
| Griffiths and Ritz, 1988 | | |
|   arable field soil | 14,743 g$^{-1}$ DW | – |
|   permanent grassland soil | 1133 g$^{-1}$ DW | – |
|   barley stubble field soil | 5547 g$^{-1}$ DW | – |

[a] Mean maximum range.
[b] Ranges of individual species' populations.
[c] Active (non-encysted) only.
[d] Humus layer only; population increase over 4 days in response to a rainfall event; peak biomass figure given.
[e] Mean value, range in parentheses.

although the number of amoebae present altered with depth. Data for naked amoebae are scarce, but Schönborn (1986a) found greater numbers (dominated by *Dactylosphaerium*, *Thecamoeba* and *Vahlkampfia* spp.) in the needle litter layers (L,F) of a spruce forest soil than in the underlying humus (H) layer. The patterns of species composition and diversity of myxomycete communities living on leaf litters and tree bark surfaces show some differentiation according to microhabitat types; these differences were mainly associated with microclimatic factors (Feest, 1987; Stephenson, 1989).

As regards humus type and content, Stout (1968) reported that Testacea were best represented in mor soils. Schönborn (1982, 1986a) recorded larger testacean populations and production rates in 'raw' (spruce forest) humus than in mull, with a larger biomass resulting from the predominance of the large Centropyxidae and Plagiopyxidae. Testacean mortality is also affected by humus type (Schönborn, 1986b). Nevertheless, abundance and production of testates can be high in some mull soils (Meisterfeld, 1986). From Testacea data compiled from numerous samples, Foissner (1987) reported significant positive correlations between both the numbers of species and individuals with the humus content of soils. A positive correlation was found for ciliates only in the case of species numbers. Foissner (1987) also compiled lists of ciliate and testate species characteristic of mull (mild, more alkaline) and moder/mor (acid) humus types (Table 2.5).

Investigations on skeletal soils and the processes of microbial colonization of sterile volcanic soils suggest that small flagellates are the first protozoan colonizers of bare sites initially without vegetation (Smith, 1974, 1985). The rate of accumulation of organic matter and the progressive

**Table 2.5.** Characteristic testacean and ciliate species of mull and mor soils (modified from Foissner, 1987).

| Humus type | Testacea | Ciliophora |
| --- | --- | --- |
| Mull | *Centropyxis constricta*<br>*Centropyxis elongata*<br>*Centropyxis plagiostoma*<br>*Geopyxella sylvicola*<br>*Paraquadrula* spp.<br>*Plagiopyxis minuta* | *Colpoda elliotti*<br>*Engelmanniella mobilis*<br>*Grossglockneria hyalini*<br>*Hemisincirra filiformis*<br>*Urosoma* spp.<br>*Urosomoida agilis* |
| Moder and mor | *Assulina* spp.<br>*Corython* spp.<br>*Nebela* spp.<br>*Plagiopyxis labiata*<br>*Schoenbornia humicola*<br>*Trigonopyxis arcula* | *Bryometopus sphagni*<br>*Dimacrocaryon*<br>  *amphileptoides*<br>*Frontonia depressa* |

establishment of suitable microhabitats appears to dictate the rate at which subsequent protozoan colonization proceeds (Smith, 1985).

There are a limited number of reports on protozoan distribution in other soil types. The protozoan species of sparsely vegetated gravel and sand deserts and semiarid woodlands tend to reflect the moisture-stressed nature of these habitats; globose Testacea are the dominant protozoa with a low diversity of ciliates, naked amoebae and small soil flagellates (Bamforth, 1984). The influential environmental factors appear to be erratic rainfall, limited soil organic matter and redistribution of litter and soil by wind or water (Bamforth, 1984; Parker *et al.*, 1984). In contrast, the proto-zoa of tropical rain forests are subject to the scouring effects of frequent heavy rainfall. Bamforth (1988) reported a larger protozoan abundance and diversity in epiphyte-associated 'suspended' soils in the tree canopy than in soils at ground level. In the rain forests, the dominant forms amongst amoebae, ciliates and flagellates were found to be representatives of *Eugly-pha* and *Nebela*, the *Colpodida* and *Mastigamoeba*, respectively (Bamforth, 1989).

## Moss peat soils

The characteristic communities of testate amoebae in habitats dominated by bryophytes and other cryptogams have been the subject of numerous studies (e.g. Heal, 1962, 1964; Meisterfeld, 1977, 1979; Smith, 1978; Gnekow, 1981; Beyens *et al.*, 1986, 1990; Warner, 1987). Marked stratification of the testacean fauna on the *Sphagnum* plant, with *Nebela* the dominant genus, has been reported by several workers (Bonnet, 1958; Heal, 1962; Meisterfeld, 1977; Gnekow, 1981). The shells of testate amoebae are well preserved in acid peat soils and are an aid for studying the stratification of bogs and in micropalaeontology (Beyens, 1984; Ellison and Ogden, 1987). A general distinction can be made between the testate and ciliate fauna of alkaline and acid peats, closely reflecting the moisture regime and base status of these different habitats (Heal, 1964; Stout and Heal, 1967).

Extensive studies of the distribution of the restricted protozoan fauna of Antarctic moss peat and fellfield (short-cushion moss and lichen) habitats have been carried out (Smith, 1978, 1982a,b, 1984; Bonnet, 1981; Smith and Tearle, 1985). The testate fauna of such moss sites are dominated by only one or a few species, e.g. *Assulina muscorum, Centropyxis aerophila, Corythion dubium, Phryganella acropodia* and *Trigonopyxis arcula* (Smith and Head-land, 1983; Smith, 1992). Similarly, an *Assulina muscorum-Corythion dubium* assemblage was found to characterize the driest and most acidic moss habitats in the high Arctic sites (Devon Island, North-West Territories, Canada) studied by Beyens *et al.* (1986, 1990). These authors noted that in Arctic soils, however, there was a much reduced frequency of *Phryganella acropodia*, an otherwise ubiquitous and abundant species in most other regions of the world. This finding agrees with earlier studies in which only low frequencies of this species were recorded in soils from Spitsbergen (Bonnet, 1964b) and Greenland (Stout, 1970), but cannot as yet be explained.

Rogerson (1982) suggested that large naked amoebae were important in the energy turnover of a temperate *Sphagnum* bog, but little further information about gymnamoebae in moss-dominated habitats exists at present.

## Cultivated and disturbed soils

The distinctive characteristics of agricultural soils are derived from the widespread use of monoculture, irrigation, fertilizers and different types of cultivations on these soils. Protozoan diversity and abundance have usually been found to be larger in cultivated than in uncultivated soils, probably as a response to enhanced soil moisture and fertility (e.g. Stout, 1960; Elliott and Coleman, 1977; Foissner, 1987). The stimulating effect of plant growth and root exudates on protozoa are well recognized and are discussed in Chapter 5. Long-term cultivation of a single crop appears to have little effect on the stability of the resident protozoa and microflora, as long as there is no major depletion of soil organic nutrients. In general, soil protozoan populations are thought to comprise very stable entities within a given soil type. Foissner (1987) found that the overall effects of cultivation (e.g. transformation of grassland into arable fields) on protozoan numbers and species are not very great. Table 2.4 includes data on protozoan numbers in a range of soils. Some human activities can have long-term deleterious effects on the composition and distribution of the soil protozoan fauna, e.g. the effects of herbicides, pesticides, excessive soil treatments (lime, salts, etc.), soil compaction, soil erosion, oil and chemical pollutants, radiation, acid rain, deforestation and fire (see review by Foissner, 1987). The use of soil protozoa as bioindicators of such disturbed soil ecosystems is discussed in Chapter 7.

## Subsoils and subsurface sediments

Although viable protozoa have been recorded from unsaturated subsoils, groundwater, muds of subterranean caves and subsurface clay sediments down to 6 m depth (Sinclair and Ghiorse, 1987), little is known of the nature of the protozoan community in these and deeper subsurface zones (arbitrarily, below 5 m depth). Several recent studies have shown that bacterial activity is widespread in aquifer sediments 10–50 m below surface and it is generally recognized that prokaryotes dominate the microbial communities of both shallow and deep subsurface sites (Ghiorse and Wilson, 1988; Sinclair and Ghiorse, 1989). It has been suggested that bacterial predation by normally 'dormant' protozoan populations may be important in aquifers contaminated with organic pollutants (Harvey and George, 1987). Sinclair and Ghiorse (1989) examined deep subsurface sediments down to 260 m depth from a site in South Carolina and found that flagellates and amoebae were distributed widely throughout the whole profile, but no ciliates were observed. Although the relative sizes of these protozoan populations were not estimated, total protozoan densities estimated by the MPN method were generally several orders of magnitude lower than those of bacterial populations. One sample from a depth of 58.8 m, however, contained large populations ($>10^3$ $g^{-1}$ DW sediment) of protozoa and viable bacteria ($>10^7$ $g^{-1}$ DW sediment). Most of the protozoa in this sample were flagellates. The general trend observed in these samples was that microbial populations generally declined with depth within the surface soil layers (c. 0–2 m), but there was no general decline in density further down. Of all the site characteristics examined, sediment texture showed the best correlation with microbial activity and abundance.

Earlier studies of subsurface sediments down to c. 8 m depth had also noted an absence of ciliates, in this case below a depth of 0.5 m in a groundwater study site in Oklahoma (Sinclair and Ghiorse, 1987, Beloin et al., 1988). It was observed that protozoan numbers declined drastically with depth to 28 $g^{-1}$ DW sediment in the unsaturated zone above the water table (at 3 m depth). The deeper layers, with varying sediment textures, contained variable numbers of protozoa; none were detected in the saturated zone at 3.6–5.2 m depth, but low numbers (0.2–67 $g^{-1}$ DW sediment) of amoebae and flagellates, probably Bodo and Cercomonas spp., were found intermittently in the layers above and below this depth.

Protozoan densities ranging from $10^4$ to $10^5$ $g^{-1}$ DW sediment were recently found in a study of subsurface aquifer sediments in the USA, but the distribution of the dominant protozoa (flagellates, 2–3 μm diameter) was not correlated with the concentration of dissolved organic carbon or the size of bacterial populations (Kinner et al., 1992).

## Geographical Distribution Patterns

All the major groups of soil protozoa contain species that are found worldwide (cosmopolites) or in virtually all major habitats (ubiquits). The common soil amoeba Naegleria, for example, shows an ubiquitous distribution (Marciano-Cabral, 1988) in common with the cosmopolitan soil flagellates Bodo, Cercomonas and Heteromita spp. (Foissner, 1991). The highly distinctive flagellate Rhyncomonas nasuta has been recorded from soil, freshwater, brackish water, seawater and even from the gut of the cockroach (Swale, 1973). Evidence is emerging, however, that many other protozoa are restricted in their distribution and that cosmopolitanism amongst soil protozoa may not be as widespread as was formerly believed.

The nature of the soil protozoan community in any given location is broadly characterized by the zonal climatic regime (e.g. polar, temperate, etc.) and the type of ecosystem (e.g. desert, woodland, bog, etc.). The type of zonal vegetation may also influence protozoan distribution; the species richness of dictyostelid myxomycetes in Japanese forest soils, for example, is relatively large and species' distribution patterns correspond closely to those of higher plant species (Cavender and Kawabe, 1990). As is the usual case with floral and faunal taxa, numbers of protozoan species decrease with increasing latitude (Smith, 1982b; Beyens *et al.*, 1986) and similar trends are evident with increasing altitude and decreasing size of habitat (Foissner, 1987).

Amongst the Testacea, there appears to be a well-defined group of genera that only occur in soil and moss (Bonnet, 1964a; Stout and Heal, 1967). Similarly, recent detailed analysis of soil ciliate communities from around the world have suggested that there are a large number of endemic ciliates (Foissner, 1987). Only about 13% of ciliate species are common to both soil and freshwater in the same geographical region and the most frequent species in soil and freshwater communities are different (Foissner, 1987). Over 50 ciliate genera are now considered to be restricted to soil and moss (Table 2.6). An interesting example is *Grossglockneria acuta* (Petz *et al.*, 1985; Aescht *et al.*, 1991). To date, no members of the family Grossglockner-idae have been found in aquatic habitats, despite examination of numerous samples (Foissner, 1987).

The most comprehensive data available on the frequency distribution of protozoa in world soils are those for Testacea. For most other groups of soil protozoa, there is too little information to permit detailed comparisons. At least two major biogeographical zones are recognized for testate amoebae: a Northern 'Laurasian' zone (North America, Europe and Asia) and a southern 'Gondwanaland' zone (South America, Africa, Australia, New Zealand and Antarctica). There appear to be significantly fewer species in the Northern Hemisphere and Bonnet (1983) suggested that this may be partly due to the prevailing equatorial winds preventing the spread of airborne testacean cysts from the Southern Hemisphere, as well as to past extensive glacial activity and marine encroachments in Europe.

Smith and Wilkinson (1986) have suggested that only comprehensive surveys that yield asymptotic curves of number of species present plotted against number of samples for each site should be used for analysis of geographical distribution. Wilkinson (1990a) warned against relying solely on the simple inspection of matrices of presence or absence of species and suggested that multivariate statistical methods, e.g. ordination analysis, should also be employed. This author analysed the distribution data of *Nebela* spp. in the Southern Hemisphere by Q-Mode and R-Mode analyses

**Table 2.6.** Genera of ciliates which are known from soil or moss only (from Foissner, 1987 and personal communication).

| | |
|---|---|
| *Amphisiellides* | *Microdiaphanosoma* |
| *Apocolpoda* | *Nivaliella* |
| *Avestina* | *Notoxoma* |
| *Balantidioides* | *Orthoamphisiella* |
| *Bardeliella* | *Orthokreyella* |
| *Bresslaua* | *Parabryophrya* |
| *Bresslauides* | *Paragastrostyla* |
| *Cirrophrya* | *Parakahliella* |
| *Corallocolpoda* | *Paramphisiella* |
| *Corticocolpoda* | *Periholosticha* |
| *Cosmocolpoda* | *Phacodinium* |
| *Dimacrocaryon* | *Platyophryides* |
| *Enchelyotricha* | *Protospathidium* |
| *Erniella* | *Pseudocyrtolophosis* |
| *Grandoria* | *Pseudoholophrya* |
| *Grossglockneria* | *Pseudokreyella* |
| *Hausmanniella* | *Pseudoplatyophrya* |
| *Hemiamphisiella* | *Pseudouroleptus* |
| *Hemisincirra* | *Reticulowoodruffia* |
| *Idiocolpoda* | *Rostrophryides* |
| *Ilsiella* | *Semiplatyophrya* |
| *Jaroschia* | *Sorogena* |
| *Kalometopia* | *Stammeridium* |
| *Krassniggia* | *Tectohymena* |
| *Kuehneltiella* | *Trihymena* |
| *Lamtostyla* | *Uroleptoides* |

(Birks, 1987), which respectively group different sites on the basis of similar groups of species or analyse similarities between the distribution patterns of different species. Of the 38 species of *Nebela* from 14 locations south of 40°S from South America, Antarctica and Australasia, 18 were restricted to the Southern Hemisphere or Gondwanaland islands and 14 were endemic to single sites (Smith and Wilkinson, 1986). Species richness decreased as latitude increased and was positively correlated with mean January temperature at each site. Some anomalies to this general pattern were observed and reflect the influence of other factors beside latitude, namely the degree of isolation or previous history of the land mass. Smith and Wilkinson (1986) suggested that the more cosmopolitan species of *Nebela* represented the more ancestral group that evolved before the Jurassic (190–140 million years BP) and that the species endemic to Gondwanaland evolved more recently. Wilkinson (1990b) also concluded that because the *Nebela* fauna of the island of South Georgia lacks any endemic species and appears to be a subset of that from neighbouring southern Chile, colonization of the island by *Nebela* must have occurred by means of windborne cysts. By contrast, the more remote and recent volcanic Marion Island has three *Nebela* spp. (*N. antarctica* and two varieties of *N. playfairi*) that probably evolved during the Quaternary and may be endemic (Wilkinson, 1990b). Out of the total of 26 testate rhizopod species that have been recorded from the continental Antarctic zone, only *N. bohemica* var. *adelia*, from Terre Adélie is thought at present to be endemic to this zone (Decloître, 1964; Smith, 1992). The overall pattern of distribution of *Nebela* spp. in the Southern Hemisphere appears to be influenced by the geological history and present geographical location of the site as well as by the restrictions imposed by harsh environmental conditions.

Recently, a remarkable ciliophoran-like flagellate, *Hemimastix amphikineta*, was discovered and for which a new phylum (Hemimastigophora) was proposed (Foissner *et al.*, 1988). Although occuring in soils from Australia and Chile, it was absent from over 1000 samples from Laurasian zones and appears to have a restricted Gondwanian distribution. Subsequent reports of further hemimastigophoran flagellates from other localities now suggest that the Hemimastigophora are more widespread and diverse than previously supposed (Foissner and Foissner, 1992). This perhaps indicates the remaining scope for discovery of new protozoan taxa and is a reminder that many regions have yet to be sampled as intensively as the Northern Temperate zone. How many species of soil protozoa are cosmopolitan and how many are restricted in their distribution will only be reliably established when such a sampling programme has been undertaken. Similarly, a better understanding of protozoan phylogeny is essential, if present-day distribution patterns are to be explained in terms of past speciation. Progress in this area is hampered by the difficulties of defining 'species' amongst asexual protozoa. Large-scale genetic diversity does, nevertheless, occur in asexual protozoa as demonstrated, for example, in natural populations of the soil amoeba *Acanthamoeba polyphaga* (Jacobson and Band, 1987). The application of modern molecular biological techniques of DNA analysis to characterize protozoan taxa by identifying stable genetic differences is likely to provide a firmer basis for the unequivocal recognition of protozoan species.

## Conclusions

Drought, waterlogging, extremes of soil temperature and atmosphere, food shortage, etc. must all be tolerated by terrestrial protozoa. The more successful species will be those that are either able to continue their activity whilst tolerating severe conditions or those that show a rapid response to changing environmental conditions. There may be an apparently paradoxical combination of a high degree of food selectivity, giving a competitive edge when food is abundant, with the ability to utilize a variety of microbial food to increase chances of survival during famine. The most successful soil protozoa, in addition to having many of these attributes, will also be widely dispersed and be able to colonize a wide range of terrestrial habitats.

We are far from achieving an overall view of protozoan distribution patterns, but some trends are emerging that suggest the influence of biogeographical factors in some cases. Further global surveys of protozoa other than the Testacea are needed to provide a more balanced picture, with more emphasis on analysis with modern multivariate methods (Ghabbour, 1991). Investigation of protozoan colonization processes might also provide useful insights. The application of molecular biological techniques to distinguish differences between related taxa should improve understanding of protistan systematics, the nature of speciation, phylogeny and present-day geographical distributions. There is as yet, however, no easily applicable method to resolve the issue of how to define species within the Protista.

Perhaps one of the greatest needs is for autecological studies, especially in hitherto neglected groups, such as the common heterotrophic flagellates (e.g. cercomonads, kinetoplastids, chrysomonads and euglenoids) as well as the gymnamoebae and the more unusual endemic species. The physiology of many of these ubiquitous or unique protozoa has barely been studied.

To know the reasons why some protozoa are more widely distributed, better adapted or more specialized than others is becoming increasingly important. Soil protozoa are now recognized for their central role as bacterial consumers (Clarholm, 1981), in improving nitrogen and sulfur mineralization (Clarholm, 1985a, 1985b; Griffiths, 1989; Gupta and Germida, 1989), as potential biological control or disease-suppressive agents (Chakraborty *et al.*, 1983), and as reservoirs of facultative microbial pathogens, such as *Legionella* (Harf and Monteil, 1988). Furthermore, species near the limits of their range of distribution are likely to show an altered distribution in response to $CO_2$-induced warming and consequent climatic change in the future. Smith and Crook (1992) have already suggested that the biological effects of these changes could be easily monitored using common soil protozoa.

## Acknowledgement

I thank John Darbyshire for helpful discussions during the preparation of this chapter.

# References

Adolph, E. (1929) The regulation of adult body size in the protozoan *Colpoda*. *Journal of Experimental Biology* 53, 269–312.

Aescht, E. and Foissner, W. (1992) Enumerating active soil ciliates by direct counting. In: Lee, J.J. and Soldo, A.T. (eds), *Protocols in Protozoology I*. Society of Protozoologists/Allen Press, Kansas, pp. B7.1-B7.4.

Aescht, E., Foissner, W. and Mulisch, M. (1991) Ultrastructure of the mycophagous ciliate *Grossglockneria acuta* (Ciliophora, Colpodea) and phylogenetic affinities of colpodid ciliates. *European Journal of Protistology* 26, 350–364.

Alabouvette, C., Coûteaux, M.M., Old, K.M. Pussard, M., Reisinger, O. and Toutain, F. (1981) Les protozaires du sol: aspects écologiques et methodologiques. *Année Biologique* 20, 255–303.

Anderson, R.V., Elliott, E.T., McLellan, J.F., Coleman, D.C., Cole, C.V. and Hunt, H.W. (1978) Trophic interactions in soils as they affect energy and nutrient dynamics. III. Biotic interactions of bacteria, amoebae and nematodes. *Microbial Ecology* 4, 361–371.

Bamforth, S.S. (1980) Terrestrial protozoa. *Journal of Protozoology* 27, 33–36.

Bamforth, S.S. (1984) Microbial distributions in Arizona deserts and woodlands. *Soil Biology and Biochemistry* 16, 133–137.

Bamforth, S.S. (1985) Symposium on 'Protozoan Ecology': the role of protozoa in litter and soils. *Journal of Protozoology* 32, 404–409.

Bamforth, S.S. (1988) Interactions between protozoa and other organisms. *Agriculture, Ecosystems and Environment* 24, 229–234.

Bamforth, S.S. (1989) Distribution of protozoa in a tropical rainforest. *Journal of Protozoology* 36, (Supplement 9A) Abstr. 53.

Bamforth, S.S. (1991) Enumeration of soil ciliate active forms and cysts by a direct count method. *Agriculture, Ecosystems and Environment* 34, 209–212.

Band, R.N. (1984) Distribution and growth of soil amoebae in a northern hardwood forest. *Journal of Protozoology* 31, (Supplement 2A) Abstr. 3.

Belcher, J.H. and Swale, E.M.F. (1976) *Spumella elongata* (Stokes) nov. comb., a colourless flagellate from soil. *Archiv für Protistenkunde* 118, 215–220.

Beloin, R.M., Sinclair, J.L. and Ghiorse, W.C. (1988) Distribution and activity of microorganisms in subsurface sediments of a pristine study site in Oklahoma. *Microbial Ecology* 16, 85–95.

Berthelin, J. and Toutain, F. (1982) Soil biology. In: Bonneau, M. and Souchier, B. (eds), *Constituents and Properties of Soils*. Academic Press, London, pp. 140–183.

Beyens, L. (1984) A concise survey of testate amoebae analysis. *Bulletin de la Societé belge de Géologie* 93, 261–266.

Beyens, L., Chardez, D. and Landtsheer, R.D. (1986) Testate amoebae populations from moss and lichen habitats in the Arctic. *Polar Biology* 5, 165–173.

Beyens, L., Chardez, D. and De Baere, D. (1990) Ecology of testate amoebae assemblages from coastal lowlands on Devon Island (NWT, Canadian Arctic). *Polar Biology* 10, 431–440.

Birks, H.J.B. (1987) Recent methodological developments in quantitative descriptive biogeography. *Annales Zoologica Fennici* 24, 165–178.

Blanton, R.L. (1990) Phylum Acrasea. In: Margulis, L., Corliss, J.O., Melkonian, M. and Chapman, D.J. (eds). *Handbook of Protoctista*. Jones and Bartlett, Boston, pp. 75–87.

Blanton, R.L. and Olive, L.S. (1983) Ultrastructure of aerial stalk formation by the ciliated protozoan *Sorogena stoianovitchae*. *Protoplasma* 116, 125–135.

Bonnet, L. (1958) Les thécamoebiens des Boullouses. *Bulletin de la Societé d'Histoire Naturelle Toulouse* 93, 529–543.

Bonnet, L. (1961) Caractères généraux des populations thécamoebiens endogées. *Pedobiologia* 1, 6–24.

Bonnet, L. (1964a) Le peuplement thécamoebiens des sols. *Revue d'Écologie et de Biologie du Sol* 1, 123–408.

Bonnet, L. (1964b) Sur le peuplement Thécamoebien de quelques sols du Spitzberg. *Bulletin de la Societé d'Histoire Naturelle Toulouse* 100, 281–293.

Bonnet, L. (1975) Types morphologiques, écologie et évolution de la thèque chez les thécamoebiens. *Protistologica* 11, 363–378.

Bonnet, L. (1981) Thécamoebiens (Rhizopoda Testacea). In: Travé, J. (ed.), Biologie des Sols. *Comité National Français des Recherches Antarctiques* 48, 23–32.

Bonnet, L. (1983) Intérêt biogéographique et paleogéographique des thécamoebiens des sols. *Annales de la Station biologique Besse-en-Chandesse* 17, 298–334.

Brooker, B.E. (1971) Fine structure of *Bodo saltans* and *Bodo caudatus* (Zoomastigophora: Protozoa) and their affinities with the Trypanosomatidae. *Bulletin of the British Museum of Natural History (Zoology)* 22, 87–102.

Buitkamp, U. (1979) Vergleichende Untersuchungen zur Temperaturadaptation von Bodenciliaten aus klimatisch verschiedenen Regionen. *Pedobiologia* 19, 221–236.

Cairns, J. and Ruthven, J.A. (1972) A test of the cosmopolitan distribution of fresh-water protozoans. *Hydrobiologia* 39, 405–427.

Carey, P.G. (1991) *Marine Interstitial Ciliates*. Chapman and Hall, London.

Casida, L.E. (1989) Protozoan response to the addition of bacterial predators and other bacteria in soil. *Applied and Environmental Microbiology* 55, 1857–1859.

Cavender, J.C. and Kawabe, K. (1990) Cellular slime molds of Japan. I. Distribution and biogeographical considerations. *Mycologia* 81, 683–691.

Chakraborty, S. and Old, K.M. (1982) Mycophagous soil amoebae: interactions with three plant pathogenic fungi. *Soil Biology and Biochemistry* 14, 247–255.

Chakraborty, S., Old, K.M. and Warcup, J.H. (1983) Amoebae from a take-all suppressive soil which feed on *Gaeumannomyces graminis tritici* and other soil fungi. *Soil Biology and Biochemistry* 15, 17–24.

Chardez, D. (1964) Sur la répartition verticale des Thécamoebiens endogés. *Bulletin de l'Institute Agronomique Stationes Recherches Gembloux* 32, 26–32.

Chardez, D. (1968) Études statistiques sur l'écologie et la morphologie des thécamoebiens (Protozoa, Rhizopoda Testacea). *Hydrobiologia* 32, 271–287.

Chardez, D. (1986) Dissemination des thécamoebiens par les arthropodes. *Les Naturalistes Belges* 67, 1–3.

Chardez, D. and Lambert, J. (1981) Thécamoebiens indicateurs biologiques (Protozoa, Rhizopoda Testacea). *Bulletin de l'Institute Agronomique Stationes Recherches Gembloux* 6, 181–203.

Clarholm, M. (1981) Protozoan grazing of bacteria in soil – impact and importance. *Microbial Ecology* 7, 343–350.

Clarholm, M. (1985a) Interactions of bacteria, protozoa and plants leading to mineralization of soil nitrogen. *Soil Biology and Biochemistry* 17, 181–188.

Clarholm, M. (1985b) Possible role for roots, bacteria, protozoa and fungi in supplying nitrogen to plants. In: Fitter, A.H., Atkinson, D., Read, D.J. and Usher, M.B. (eds), *Ecological Interactions in Soil: Plants, Microbes, and Animals*. Blackwell Scientific Publications, Oxford, pp. 355–365.

Corliss, J.O. and Esser, S.C. (1974) Comments on the role of the cyst in the life cycle and survival of free-living protozoa. *Transactions of the American Microscopial Society* 93, 578–593.

Coûteaux, M.M. (1967) Une technique d'observation des thécamoebiens du sol pour l'estimation de leur densité absolue. *Revue d'Écologie et de Biologie du Sol* 4, 593–596.

Coûteaux, M.M. (1972) Distribution des thécamoebiens de la litière et de l'humus de deux sols forestier d'humus brut. *Pedobiologia* 12, 237–243.

Coûteaux, M.M. (1975) Estimation quantitative des thécamoebiens édaphiques par rapport à la surface du sol. *Compte rendu hebdomadaires des séances de l'Académie des Sciences, Paris* 281, 739–741.

Coûteaux, M.M. (1977) Reconstitution d'une nouvelle communité thécamoebienne dans la litière d'une fôret incendiée en region submediterranéene. *Ecological Bulletins (Stockholm)* 25, 102–108.

Coûteaux, M.M. (1985) Relationships between testate amoebae and fungi in humus microcosms. *Soil Biology and Biochemistry* 17, 339–345.

Coûteaux, M.M. and Devaux, J. (1983) Effet d'un enrichissement en champignons sur la dynamique d'un peuplement thécamoebien d'un humus. *Revue d'Écologie et de Biologie du Sol* 20, 519–545.

Coûteaux, M.M. and Ogden, C.G. (1988) The growth of *Tracheleuglypha dentata* (Rhizopoda: Testacea) in clonal cultures under different trophic conditions. *Microbial Ecology* 15, 81–93.

Cowling, A.J. (1986) Culture methods and observations of *Corythion dubium* and *Euglypha rotunda* (Protozoa, Rhizopoda) isolated from maritime Antarctic moss peats. *Protistologica* 22, 181–191.

Cowling, A.J. and Smith, H.G. (1988) Protozoa in the microbial communities of maritime Antarctic fellfields. Deuxiéme Colloque sur les Écosystèmes Terrestres Subantarctiques, 1986, Paimpont. *Comité National Français des Recherches Antarctiques* 58, 205–213.

Cutler, D.W. and Crump, L.M. (1927) The qualitative and quantitative effects of food on the growth of a soil amoeba (*Hartmanella hyalina*). *Journal of Experimental Biology* 5, 155–165.

Cutler, D.W., Crump, L.M. and Sandon, H. (1922) A quantitative investigation of the bacterial and protozoan population of the soil, with an account of the protozoan fauna. *Philosophical Transactions of the Royal Society of London Series B* 211, 317–350.

Darbyshire, J.F., Wheatley, R.E., Greaves, M.P. and Inkson, R.H.E. (1974) A rapid method for estimating bacterial and protozoan populations in soil. *Revue d'Écologie et de Biologie du Sol* 11, 465–475.

Darbyshire, J.F., Elston, D.A., Simpson, A.E.F., Robertson, M.D. and Seaton, A. (1992) Motility of a common soil flagellate *Cercomonas* sp. in the presence of aqueous infusions of fungal spores. *Soil Biology and Biochemistry* 24, 827–831.

Davis, R.C. (1980) Structure and function of two Antarctic terrestrial moss communities. *Ecological Monographs* 51, 125–143.

Decloître, L. (1960) Perfectionements apportés à un appareil pour une technique d'isolement des microorganismes du sol, des mousses, des eaux. *Internationale Revue de Hydrobiologie et Hydrographie* 45, 169–171.

Decloître, L. (1964) Thécamoebiens de la XIIème Expédition Antarctique Française. *Publications de l'Expédition polair français* 259, 47pp.

Ebringer, L., Balan, J. and Nemec, P. (1964) Incidence of antiprotozoal substances in Aspergillaceae. *Journal of Protozoology* 11, 153–156.

Elliott, E.T. and Coleman, D.C. (1977) Soil protozoan dynamics in a shortgrass prairie. *Soil Biology and Biochemistry* 9, 113–118.

Elliott, E.T., Coleman, D.C. and Cole, C.V. (1980) Habitable pore space and microbial trophic interactions. *Oikos* 35, 327–335.

Ellison, R.L. and Ogden, C.G. (1987) A guide to the study and identification of fossil testate amoebae in Quaternary lake sediments. *Internationale Revue gesamte Hydrobiol.* 72, 639–652.

Epstein, S.S. and Shiaris, M.P. (1992) Size-selective grazing of coastal bacterioplankton by natural assemblages of pigmented flagellates, colorless flagellates, and ciliates. *Microbial Ecology* 23, 211–225.

Feest, A. (1987) The quantitive ecology of soil mycetozoa. *Progress in Protistology* 2, 331–361.

Fenchel, T. (1974) Intrinsic rate of natural increase: the relationship with body size. *Oecologia* 14, 317–326.

Foissner, I. and Foissner, W. (1992) The fine structure of two new hemimastigophoran flagellates related to *Spironema multiciliatum*, Klebs 1893. *European Journal of Protistology* 28, (Supplement 340) Abstr. 46.

Foissner, W.F. (1980) Colpodide Cilaten (Protozoa: Ciliophora) aus alpinen Böden. *Zoologische Jahrbücher, Systematik, Okologie und Geographie der Tiere* 107, 391–432.

Foissner, W. (1987) Soil protozoa: fundamental problems, ecological significance, adaptations in ciliates and testaceans, bioindicators and guide to the literature. *Progress in Protistology* 2, 69–212.

Foissner, W. (1991) Diversity and ecology of soil flagellates. In: Patterson, D.J. and Larsen, J. (eds), *The Biology of Free-living Heterotrophic Flagellates.* Systematics Association Special Volume No. 45, Clarendon Press and The Systematics Association, Oxford, pp. 93–112.

Foissner, W., Blatterer, H. and Foissner, I. (1988) The Hemimastigophora (*Hemimastix amphikineta* nov. gen., nov. spec.), a new protistan phylum from Gondwanian soils. *European Journal of Protistology* 23, 361–383.

Ghabbour, S.I. (1991) Towards a zoosociology of the soil fauna. *Revue d'Écologie et de Biologie du Sol* 28, 77–90.

Ghiorse, W.C. and Wilson, J.T. (1988) Microbiology of the terrestrial subsurface. *Advances in Applied Microbiology* 33, 107–172.

Gnekow, M.A. (1981) Beobachtungen zur Biologie und Ultrastruktur der moosbewohnenden Thekamöbe *Nebela tincta* (Rhizopoda). *Archiv für Protistenkunde* 124, 36–69.

Golemansky, V. (1977) Adaptations morphologiques des thécamoebiens psammobiontes du psammal supralittoral des mers. *Acta Protozoologica* 17, 141–152.

Gonzalez, J.M., Sherr, E.B. and Sherr, B.F. (1990) Size-selective grazing on bacteria by natural assemblages of estuarine flagellates and ciliates. *Applied and Environmental Microbiology* 56, 583–589.

Griffiths, B.S. (1989) Enhanced nitrification in the presence of bacteriophagous protozoa. *Soil Biology and Biochemistry* 21, 1045–1051.

Griffiths, B.S. (1990) A comparison of microbial feeding nematodes and protozoa in the rhizosphere of different plants. *Biology and Fertility of Soils* 9, 83–88.

Griffiths, B.S. and Ritz, K. (1988) A technique to extract, enumerate and measure protozoa from mineral soils. *Soil Biology and Biochemistry* 20, 163–173.

Grospietsch, T. (1965) Schalenamöben im Boden. *Mikrokosmos* 54, 14–18.

Gupta, V.V.S.R. and Germida, J.J. (1989) Influence of bacterial–amoebal interactions on sulfur transformations in soil. *Soil Biology and Biochemistry* 21, 921–930.

Harf, C. and Monteil, H. (1988) Interactions between free-living amoebae and *Legionella* in the environment. In: Jenkins, D. and Olsen, B.J. (eds), *Water and Wastewater Microbiology.* Pergamon Press, Oxford, pp. 235–239.

Harvey, R.W. and George, L.H. (1987) Growth determination for unattached bacteria in a contaminated aquifer. *Applied and Environmental Microbiology* 53, 2992–2996.

Have, A. (1987) Experimental island biogeography: immigration and extinction of ciliates in microcosms. *Oikos* 50, 218–224.

Heal, O.W. (1962) The abundance and micro-distribution of testate amoeba (Rhizopoda, Testacea) in *Sphagnum. Oikos* 13, 35–47.

Heal, O.W. (1963a) Soil fungi as food for amoebae. In: Doeksen, J. and Van der Drift, J. (eds), *Soil Organisms.* North-Holland Publishing Co., Amsterdam, pp. 289–297.

Heal, O.W. (1963b) Morphological variation in certain Testacea (Protozoa: Rhizopoda). *Archiv für Protistenkunde* 106, 351–368.

Heal, O.W. (1964) Observations on the seasonal and spatial distribution of Testacea (Protozoa: Rhizopoda) in *Sphagnum. Journal of Animal Ecology* 33, 395–412.

Heal, O.W. and Felton, M.J. (1970) Soil amoebae: their food and their reaction to microflora exudates. In: Watson, A. (ed.), *Animal Populations in Relation to their Food Resources.* Blackwell Scientific Publications, Oxford, pp. 145–162.

Hekman, W.E., Van den Boogert, P.H.J.F. and Zwart, K.B. (1992) The physiological ecology of a novel obligate mycophagous flagellate. *FEMS Microbiology Ecology* 86, 255–265.

Hughes, J. and Smith, H.G. (1989) Temperature relations of *Heteromita globosa* Stein in Signy Island fellfields. In: Heywood, R.B. (ed.), *University Research in Antarctica. Proceedings of the British Antarctic Survey Special Topic Award Scheme Symposium*, 9–10 November 1988. British Antarctic Survey, Cambridge, pp. 117–122.

Jacobson, L.M. and Band, R.N. (1987) Genetic heterogeneity in a natural population of *Acanthamoeba polyphaga* from soil, and isoenzyme analysis. *Journal of Protozoology* 34, 83–86.

Kinner, N.E., Harvey, R.W., Bunn, A.L. and Warren, A. (1992) Investigation of Protozoa in an organically contaminated subsurface environment. Presented at the *6th International Symposium on Microbial Ecology*, September 1992, Barcelona (abstr.), p. 241.

Kuikman, P.J., Van Elsas, J.D., Jansen, A.G., Burgers, S.L.G.E. and Van Veen, J.A. (1990) Population dynamics and activity of bacteria and protozoa in relation to their spatial distribution in soil. *Soil Biology and Biochemistry* 22, 1063–1073.

Laminger, H. (1978) The effects of soil moisture on the testacean species *Trinema enchelys* (Ehrenberg) Leidy in a high mountain brown earth podsol and its feeding behaviour. *Archiv für Protistenkunde* 120, 446–454.

Laminger, H. and Bucher, M. (1984) Feeding behaviour of some terrestrial Testacea (Protozoa, Rhizopoda). *Pedobiologia* 27, 313–322.

Laminger, H. and Sturn, R. (1984) Influence of nutrition on encystment and excystment of Testacea (Protozoa, Rhizopoda). *Pedobiologia* 27, 241–244.

Lee, J.J. and Soldo, A.T. (eds) (1992) *Protocols in Protozoology I.* Society of Protozoologists/Allen Press, Kansas.

Leedale, G.F. (1985) Euglenida, Butschli 1884. In: Lee, J.J., Hutner, S.H. and Bovee, E.C. (eds), *An Illustrated Guide to the Protozoa .* Society of Protozoologists, Lawrence, Kansas, pp. 41–54.

Levrat, P., Alabouvette, C. and Pussard, M. (1987) Effect of predation by protozoa on the interactions between bacteria and fungi. *Revue d'Écologie et de Biologie du Sol* 24, 503–514.

Levrat, P., Pussard, M., Steinberg, C. and Alabouvette, C. (1991) Regulation of

*Fusarium oxysporum* populations introduced into soil: the amoebal predation hypothesis. *FEMS Microbial Ecology* 86, 123–130.

Lousier, J.D. (1982) Colonization of decomposing deciduous leaf litter by Testacea (Protozoa, Rhizopoda): species succession, abundance and biomass. *Oecologia* 52, 381–388.

Lousier, J.D. and Parkinson, D. (1981) Evaluation of a membrane filter technique to count soil and litter testacea. *Soil Biology and Biochemistry* 13, 209–215.

Lousier, J.D. and Parkinson, D. (1984) Annual population dynamics and production ecology of Testacea (Protozoa: Rhizopoda) in an aspen woodland soil. *Soil Biology and Biochemistry* 16, 103–114.

Lüftenegger, G. and Foissner, W. (1991) Morphology and biometry of twelve soil testate amoebae (Protozoa, Rhizopoda) from Australia, Africa and Austria. *Bulletin of the British Museum of Natural History (Zoology)* 57, 1–16.

Lüftenegger, G., Foissner, W. and Adam, H. (1985) *r*– and *K*-selection in soil ciliates: a field and experimental approach. *Oecologia* 66, 574–579.

Lüftenegger, G., Petz, W., Berger, H., Foissner, W. and Adam, H. (1988) Morphologic and biometric characterization of twenty-four soil testate amoebae (Protozoa, Rhizopoda). *Archiv für Protistenkunde* 136, 153–189.

Maguire, B. (1963) The exclusion of *Colpoda* (Ciliata) from superficially favourable habitats. *Ecology* 44, 781–784.

Marciano-Cabral, F. (1988) Biology of *Naegleria* spp. *Microbiological Reviews* 52, 114–133.

Meisterfeld, R. (1977) Die horizontale und vertikale Verteilung der Testaceen (Rhizopoda, Testacea) in *Sphagnum*. *Archiv für Hydrobiologia* 79, 319–356.

Meisterfeld, R. (1979) Clusteranalytische Differenzierung der Testaceenzönosen (Rhizopoda, Testacea) in *Sphagnum*. *Archiv für Protistenkunde* 121, 270–307.

Meisterfeld, R. (1986) The importance of protozoa in a beech forest ecosystem. *Advances in Protozoological Research* 33, 291–299.

Meisterfeld, R. (1989) Die Bedeutung der protozoen im Kohlenstoffhaushalt eines Kalkbuchenwaldes (Zur Funktion der Fauna in einem Mullbuchenwald 3). *Verhandlungen der Gesellschaft für Ökologie* 17, 221–227.

Mitchell, G.C., Baker, J.H. and Sleigh, M.A. (1988) Feeding of a freshwater flagellate, *Bodo saltans*, on diverse bacteria. *Journal of Protozoology* 35, 219–222.

Napolitano, A. (1983) Presence of amoebae in the rhizosphere of a beach grass. *Journal of Protozoology* 30, 540–541.

Nisbet, B. (1984) *Nutrition and Feeding Strategies of Protozoa*. Croom Helm, London.

Ogden. C.G. (1983) Observations on the systematics of the genus *Difflugia* in Britain (Rhizopoda, Protozoa). *Bulletin of the British Museum of Natural History* (Zoology) 44, 1–73.

Ogden, C.G. and Pitta, P. (1990) Biology and ultrastructure of the mycophagous soil testate amoeba *Phryganella acropodia* (Protozoa, Rhizopoda). *Biology and Fertility of Soils* 9, 101–109.

Old, K.M. and Chakraborty, S. (1986) Mycophagous soil amoebae: their biology and significance in the ecology of soil-borne plant pathogens. *Progress in Protistology* 1, 163–194.

Olive, L.S. and Blanton, R.L. (1980) Aerial sorocarp development by the aggregative ciliate *Sorogena stoianovitchae*. *Journal of Protozoology* 27, 293–299.

Page, F.C. (1976) *An Illustrated Key to Freshwater and Soil Amoebae*. Freshwater Biological Association Scientific Publicaton No. 34, Freshwater Biological Association, Ambleside.

Page, F.C. (1988) *A New Key to Freshwater and Soil Gymnamoeba*. Freshwater Biological Association, Ambleside.

Parker, L.W., Freckman, D.W., Steinberger, Y., Driggers, L. and Whitford, W.G. (1984) Effects of simulated rainfall and litter quantities on desert soil biota: soil respiration, microflora and protozoa. *Pedobiologia* 27, 185–190.

Persson, T., Bååth, E., Clarholm, M., Lundkvist, H., Söderström, B.E. and Sohlenius, B. (1980) Trophic structure, biomass dynamics and carbon metabolism of soil organisms in a Scot's pine forest. *Ecological Bulletins (Stockholm)* 32, 419–459.

Petz, W. and Foissner, W. (1988) Spatial separation of terrestrial ciliates and testaceans (Protozoa) – a contribution to soil ciliatostasis. *Acta Protozoologica* 27, 249–258.

Petz, W., Foissner, W. and Adam, H. (1985) Culture, food selection and growth rate in the mycophagous ciliate *Grossglockneria acuta* Foissner, 1980: first evidence of autochthonous soil ciliates. *Soil Biology and Biochemistry* 17, 871–875.

Pussard, M. and Delay, F. (1985) Dynamique de population d'amibes libres endogées (Amoebida, Protozoa). I. Evaluation du degré d'activité en microcosme et observations préliminaires sur la dynamique de population de quelques espèces. *Protistologica* 21, 5–15.

Rauenbusch, K. (1987) Biologie und Feinstruktur (REM-Untersuchungen) terrestrischer Testaceen in Waldböden (Rhizopoda, Protozoa). *Archiv für Protistenkunde* 134, 191–294.

Rivera, F., Roy-Acotla, G., Rosas, I., Ramirez, E., Bonilla, P. and Lares, F. (1987) Amoebae isolated from the atmosphere of Mexico City and environs. *Environmental Research* 42, 149–154.

Rogerson, A. (1982) An estimation of the annual production and energy flow of the large naked amoebae population inhabiting a *Sphagnum* bog. *Archiv für Protistenkunde* 126, 145–149.

Rogerson, A. and Berger, J. (1981) The effects of cold temperatures and crude oil on the abundance and activity of protozoa in a garden soil. *Canadian Journal of Zoology* 59, 1554–1560.

Rutherford, P.M. and Juma, N.G. (1992) Influence of texture on habitable pore space and bacterial–protozoan populations in soil. *Biology and Fertility of Soils* 12, 221–227.

Sandon, H. (1927) *The Composition and Distribution of the Protozoan Fauna of the Soil*. Oliver and Boyd, Edinburgh.

Schlichting, H.E., Speziale, B.J. and Zink, R.M. (1977) Dispersal of algae and Protozoa by Antarctic flying birds. *Antarctic Journal of the United States* 122, 14–21.

Schönborn, W. (1962) Zur Ökologie der sphagnikolen, bryokolen und terrikolen Testaceen. *Limnologica* 1, 231–254.

Schönborn, W. (1965) Untersuchungen über die Ernährung bodenbewohnender Testaceen. *Pedobiologia* 5, 205–210.

Schönborn, W. (1982) Estimation of annual production of testacea (Protozoa) in mull and moder (II) *Pedobiologia* 23, 383–393.

Schönborn, W. (1983) Beziehungen zwischen Produktion, Mortalität und Abundanz terrestrischer Testaceen-Gemeinschaften. *Pedobiologia* 25, 403–412.

Schönborn, W. (1986a) Population dynamics and production biology of testate amoebae (Rhizopoda, Testacea) in raw humus of two coniferous soils. *Archiv für Protistenkunde* 132, 325–342.

Schönborn, W. (1986b) Comparisons between the characteristics of the production of Testacea (Protozoa, Rhizopoda) in different forms of humus. *Symposia Biologica Hungarica* 33, 275–284.

Schönborn, W. (1989) The topophenetic analysis as a method to elucidate the phylogeny of testate amoebae (Protozoa, Testacealobosia and Testaceafilosia) *Archiv für Protistenkunde* 137, 223–245.

Schönborn, W. and Peschke, T. (1988) Biometric studies of species, races, ecopheno-types and individual variations of soil-inhabiting Testacea (Protozoa, Rhizopoda) including *Trigonopyxis minuta* n. sp. and *Corythion asperulum* n. sp. *Archiv für Protistenkunde* 136, 345–363.

Schönborn, W. and Peschke, T. (1990) Evolutionary studies on the *Assulina-Valkano-via* complex in *Sphagnum* and soil. *Biology and Fertility of Soils* 9, 95–100.

Schönborn, W., Foissner, W. and Meisterfeld, R. (1983) Licht- und rasterelektronen-mikroskopische untersuchungen zur schalenmorphologie und rassenbildung bodenbewohnender testaceen (Protozoa: Rhizopoda) sowie vorschläge zur biometrischen charakterisierung von testaceen-schalen. *Protistologica* 19, 553–566.

Schönborn, W., Petz, W., Wanner, M. and Foissner, W. (1987) Observations on the morphology and ecology of the soil-inhabiting testate amoebae *Schoenbornia humicola* (Schönborn, 1964) Decloître, 1964 (Protozoa, Rhizopoda). *Archiv für Protistenkunde* 134, 315–330.

Sibbald, M.J. and Allbright, L.J. (1988) Aggregated and free bacteria as food sources for heterotrophic microflagellates. *Applied and Environmental Microbiology* 54, 613–616.

Sinclair, J.L. and Ghiorse, W.C. (1987) Distribution of protozoa in subsurface sedi-ments of a pristine groundwater study site in Oklahoma. *Applied and Environ-mental Microbiology* 53, 157–163.

Sinclair, J.L. and Ghiorse, W.C. (1989) Distribution of aerobic bacteria, protozoa, algae and fungi in deep subsurface sediments. *Geomicrobiology Journal* 7, 15–31.

Singh, B.N. (1942) Selection of bacterial food by soil flagellates and amoebae. *Annals of Applied Biology* 29, 18–22.

Singh, B.N. (1945) The selection of food by soil amoebae and the toxic effects of bacterial pigments and other products on soil protozoa. *British Journal of Experimental Pathology* 26, 316–325.

Singh, B.N. (1946) A method of estimating the numbers of soil protozoa, especially amoebae, based on their differential feeding on bacteria. *Annals of Applied Biology* 33, 112–120.

Sleigh, M.A. (1989) *Protozoa and other Protists*. Arnold, London.

Smith, H.G. (1973) The temperature relations and bi-polar biogeography of the ciliate genus *Colpoda*. *Bulletin of the British Antarctic Survey* No. 37, 7–13.

Smith, H.G. (1974) The colonization of volcanic tephra on Deception Island by Protozoa. *Bulletin of the British Antarctic Survey* No. 38, 49–58.

Smith, H.G. (1978) The distribution and ecology of the Terrestrial Protozoa of sub-Antarctic and maritime Antarctic Islands. *British Antarctic Survey Scientific Reports* No. 95, 104 pp.

Smith, H.G. (1982a) A comparative study of the terrestrial Protozoa inhabiting moss turf peat on Iles Crozet, South Georgia and the South Orkney Islands. *Comité National Français des Recherches Antarctiques* 51, 137–145.

Smith, H.G. (1982b) The terrestrial protozoan fauna of South Georgia. *Polar Biology* 1, 173–179.

Smith, H.G. (1984) Protozoa of Signy Island fellfields. *Bulletin of the British Antarc-tic Survey* No. 64, 55–61.

Smith, H.G. (1985) The colonization of volcanic tephra on Deception Island by Protozoa: long-term trends. *Bulletin of the British Antarctic Survey* No. 66, 19–33.

Smith, H.G. (1992) Distribution and ecology of the testate rhizopod fauna of the continental Antarctic zone. *Polar Biology* 12, 629–634.

Smith, H.G. and Crook, M.J. (1992) Adaptations of flagellates and ciliates in Antarctic fellfields. *European Journal of Protistology* 28, (Supplement 358) Abstr. 124.

Smith, H.G. and Headland, R.K. (1983) The population ecology of soil testate rhizopods on the sub-Antarctic island of South Georgia. *Revue d'Écologie et de Biologie du Sol* 20, 269–286.

Smith, H.G. and Tearle, P.V. (1985) Aspects of microbial and protozoan abundances in Signy Island fellfields. *Bulletin of the British Antarctic Survey* No. 68, 83–90.

Smith, H.G. and Wilkinson, D.M. (1986) Biogeography of testate rhizopods in the southern temperate and Antarctic zones. Deuxième Colloque sur les Écosystèmes Terrestres Subantarctiques, 1986, Paimpont. *Comité National Français des Recherches Antarctiques* 58, 83–96.

Smith, H.G., Hughes, J. and Moore, S.J. (1990) Growth of Antarctic and temperate terrestrial Protozoa under fluctuating temperature regimes. *Antarctic Science* 2, 313–320.

Sneddon, E.M. (1983) The interaction of protozoa and bacteria in a soil column. Unpublished MSc thesis, University of Aberdeen.

Stephenson, S.L. (1989) Distribution and ecology of myxomycetes in temperate forests. II. Patterns of occurrence on bark surfaces of living trees, leaf litter and dung. *Mycologia* 81, 608–621.

Stout, J.D. (1960) Biological studies of some tussock-grassland soils XVII. Protozoa of two cultivated soils. *New Zealand Journal of Agricultural Research* 3, 237–244.

Stout, J.D. (1962) An estimation of microfaunal populations in soils and forest litter. *Journal of Soil Science* 13, 314–320.

Stout, J.D. (1963) Some observations on the Protozoa of some beechwood soils on the Chiltern Hills. *Journal of Animal Ecology* 32, 281–287.

Stout, J.D. (1968) The significance of the protozoan fauna in distinguishing mull amd mor of beech. *Pedobiologia* 8, 387–400.

Stout, J.D. (1970) The bacteria and protozoa of some soil samples from Scoresby Land, East Greenland. *Meddeleser om Grønland* 184, 1–23.

Stout, J.D. (1974) Protozoa. In: Dickinson, H. and Pugh, G.J.F. (eds), *Biology of Plant Litter Decomposition.* Academic Press, New York, pp. 383–420.

Stout, J.D. and Heal, O.W. (1967) Protozoa. In: Burges, A. and Raw, F (eds), *Soil Biology.* Academic Press, London, pp. 149–195.

Swale, E.M.F. (1973) A study of the colourless flagellate *Rhyncomonas nasuta* (Stokes) Klebs. *Biological Journal of the Linnean Society* 5, 255–264.

Tan, Y., Bond, W.J., Rovira, A.D., Brisbane, P.G. and Griffin, D.M. (1991) Movement through soil of a biological control agent, *Pseudomonas fluorescens. Soil Biology and Biochemistry* 23, 821–825.

Taylor, W.D. (1979) Overlap among cohabiting ciliates in their growth responses to various prey bacteria. *Canadian Journal of Zoology* 57, 949–951.

Taylor, W.D. and Shuter, B.J. (1981) Body size, genome size and intrinsic rate of increase in ciliated protozoa. *American Naturalist* 118, 160–172.

Volz, P. (1951) Über die Mikrofauna des Waldbodens. *Zoologische Jahrbücher, Systematik, Okologie und Geographie der Tiere* 79, 514–566.

Warner, B.G. (1987) Abundance and diversity of testate amoebae (Rhizopoda, Testacea) in *Sphagnum* peatlands in southwestern Ontario, Canada. *Archiv für Protistenkunde* 133, 173–189.

Wickham, S.A. and Lynn, D.H. (1990) Relations between growth rate, cell size, and

DNA content in colpodean ciliates (Ciliophora: Colpodea). *European Journal of Protistology* 25, 345–352.

Wilkinson, D.M. (1990a) Multivariate analysis of the biogeography of the protozoan genus *Nebela* in southern temperate and Antarctic zones. *European Journal of Protistology* 26, 117–121.

Wilkinson, D.M. (1990b) Glacial refugia in South Georgia? Protozoan evidence. *Quaternary Newsletter* 62, 12–13.

Yongue, W.H. and Cairns, J. (1978) The role of flagellates in pioneer protozoan colonization of artificial substrates. *Polskie Archiwum Hydrobiologii* 25, 787–801.

# Soil Microenvironment  3

T. Hattori

*Institute of Genetic Ecology, Tohoku University, Katahira,*
*Aoba, Sendai 980, Japan.*

## Introduction

Ecological studies require some knowledge about the surrounding environment at a scale relevant to the dimensions of the organisms involved. In microbial ecology of soils, knowledge about the environment at the microscopical scale, i.e. the soil microenvironment, has been more slowly acquired than details about the physiology of soil microorganisms. Nutrient concentration, pH, the partial pressure of oxygen and temperature of the microbial culture, for example, are well-established factors affecting many soil microorganisms cultivated in laboratory media, but the heterogenity and complexity of soil has hampered investigation of the ranges of these factors that exist in the immediate vicinity of microorganisms in soil.

In this chapter, the development of the concept of the soil microenvironment is discussed briefly, followed by a more detailed treatment of the distribution of protozoa and other microorganisms in this environment. As individual soil aggregates are convenient units of soil structure, the distribution and behaviour of soil microorganisms in soil aggregates are emphasized. Finally, some aspects of the soil environment that clearly need to be investigated in the future are discussed.

## Development of the Concept of the Soil Microenvironment

In the 19th century, protozoa and fungi were not widely accepted as active components of the soil community. Bacteria were frequently considered as the major agents involved in transforming organic substances in soil. For

43

this reason, it was to be expected that the concept of the microenvironment originated from bacteriologists. Frankel (1887) made an extensive study of bacteria in various soils by the dilution plate technique recently invented by Koch and found that bacterial cells occurred not only in surface soil, but also deeper in the soil. He believed that bacteria in surface layers of soil were transported to the deeper layers by downward percolating rain-water through small soil pores. Probably this was the first mention of the soil microenvironment of bacteria. Duclaux, who was a close colleague of Pasteur, published a large textbook in 1898 entitled *Traité de Microbiologie*. In this book, he described the concept of the soil microenvironment in which saprophytic bacteria as well as nitrifiers were present; at that time saprophytic bacteria and nitrifying bacteria were the most well-known soil microorganisms. Based on the idea that saprophytic bacteria depend on soluble organic food, he listed the following important chacteristics of the soil environment: downward percolation of water, wetting of the soil matrix, capillary forces, water holding capacity, pore space, adsorptive capacity of soil and distribution of soil organic matter. It is remarkable that he attempted to assess the overall influence of these characteristics on bacterial distri-bution and activities in soil at the time when the physico-chemical nature of soil was so imprecisely understood.

Early in the present century, protozoa and fungi became widely recog-nized as active components of the soil community thanks to the pioneering works of Russell, Hutchinson, Cutler, Crump and Waksman. During this period, conceptions of the microenvironment were simple and unsophisti-cated. Some authors, such as Cutler and Crump (1935), regarded soil as a rich habitat for microorganisms. They wrote in their book *Problems in Soil Microbiology*:

> A crumb may therefore be looked upon as an aggregation of particles held together by the colloidal material to form a larger mass, which, in many respects, behaves as a structual unit. From the method of construction it follows that such crumbs contain small spaces, many of them of only capillary size, which are conveniently known as the micro-pores, while the pores between the crumbs themselves may be called the macro-pores; all the spaces in the soil either between the crumbs, or incorporated within the crumb structure itself, allow of the presence of air and of water, and also of gaseous and liquid circulation.

Between 1945 and 1951, Quastel and his coworkers published a series of important studies of soil nitrification and proposed a new model of the soil microenvironment for nitrifying bacteria (Lees and Quastel, 1946a,b; Quas-tel and Schofield, 1951). From a kinetic analysis of nitrification, they con-cluded that nitrifying bacteria were active only at restricted sites in the soil crumb, where ammonium ions were retained by base exchange. Their model emphasized two aspects of the microenvironment; the importance of solid surfaces and the soil crumb or aggregate as a collection of solid surfaces.

McLaren (1954) suggested that many soil biochemical reactions occur on solid surfaces where the concentrations of ions are different from those of the bulk solution. McLaren and Skujins (1968) coined the terms micro-environment and molecular environment. In their view, the former consisted of dimensions that are commensurate with the size of the organism in question; their molecular environment was the immediate surface of soil particles and microorganisms themselves. McLaren predicted that the optimal pH for a biochemical reaction at a surface would differ from that in the bulk solution. Such effects were observed for chymotrypsin (McLaren and Estermann, 1957) and oxidations of organic substances by bacteria adsorbed on an anion exchange resin (Hattori and Furusaka, 1960, 1961; Hattori and Hattori, 1963). Later, McLaren and his colleagues developed a mathematical model for microbial action and growth in such a soil microenvironment (McLaren, 1971).

Clay particles were regarded as an important factor affecting soil microorganisms by several authors. Marshall (1971) showed that clay particles and bacterial cells can form a complex by the interaction between negative charges of the bacterial surface and positive charges along the edges of clay platelets. Hattori (1970) showed that ferric ions can form another type of cell–clay complex by means of the chelating ability of these ions fixed on the surface of the clay platelets. Effects of clay particles on microbial respiration, growth, spore germination and gene-transfer amongst soil microorganisms were extensively studied by Stotzky (1986). More recent investigations of the interactions between microorganisms and clay particles have been reviewed by Robert and Chenu (1992).

In 1959, Emerson proposed a model of the soil crumb or aggregate based on results of his dispersion experiments. In his model, a soil aggregate consists of domains of oriented clay and quartz particles that are linked together by organic material as well as by electrostatic forces. He considered that cultivation of the soil resulted in the breaking of some organic matter bonds joining domains or between domains and quartz particles. Any organic matter exposed by this process he thought would be liable to be attacked by microorganisms. This model was able to explain experiments of Rovira and Greacen (1957), who found an increase in microbial oxidation after shearing wet soil crumbs. Greenwood (1968) interpreted the stimulation of organic matter decomposition in aggregates either by drying, freezing, thawing, compaction, or shaking with water in terms of rearrangements of the units of the structure within Emerson's model of soil aggregates.

Edwards and Bremner (1967) showed that in mineral soils microaggregates (mostly less than 250 μm in diameter) are less easily dispersed than macroaggregates. They disrupted the interparticle bonding in microaggregates by the application of mechanical energy, such as by sonication or prolonged shaking with water. Edwards and Bremner suggested that microaggregates consist largely of clay–polyvalent metal–organic complexes. Different organic materials from both plants and microorganisms are involved in the stability of these organic complexes (Tisdall and Oades, 1982; Oades, 1984). The stability of microaggregates is also enhanced by multivalent cations that act as bridges between organic colloids and clays.

Greenwood (1968) used the concept of a soil aggregate to explain the presence of aerobic and anaerobic sites in soil crumbs at the same time. Assuming a uniform distribution of bacteria within soil aggregates, he derived an equation that defined the area within an aggregate where aerobic metabolism was possible. He assumed that anaerobic bacteria would be active in the remaining area of the aggregates. This model of anaerobiosis was extended to soil aggregates under field conditions by Smith (1980), who calculated the rates of denitrification in soil aggregates of various sizes. Tiedje and his colleagues measured the oxygen concentration within soil aggregates with oxygen microelectrodes. They also calculated the oxygen distribution using Smith's revised model and found reasonable agreement between the measured and calculated anaerobic radii (Tiedje et al., 1984).

The pore size distribution of soil samples is another important aspect of the soil microenvironment. Wallace (1956, 1958a, 1958b) showed that nematodes moved mainly in pore spaces between soil aggregates when water was present in pores larger than their own dimensions. Hattori (1963, 1973) interpreted effects of sonic dispersion on the distribution of bacterial cells in soil samples by assuming that most soil bacteria other than spores inhabit the capillary pores where water is held strongly. Griffin (1963, 1972) studied the effect of soil moisture on the growth of soil fungi in terms of matric potential and emphasized the relationship between the matric potential and the pore size distribution in a soil sample. Cook and Papendick (1970) showed that the response of fungi to water content in soils with different textures can be described similarly in terms of the matric potential, although the response occurred at different water contents in different soils. Recently, similar studies have been made with protozoa (Darbyshire, 1976; Elliott et al., 1980; Vargas and Hattori, 1986; Kuikman et al., 1989; Heijnen et al., 1990; Postma et al., 1990).

Microbial feeding activities in soils are closely related to the habitable pore space of different microorganisms (Elliott et al., 1980). Elliott and Coleman (1988) conceived a hierarchical series of soil aggregates of different sizes to explain many biochemical changes in soil and ecological relationships between organisms in detrital food webs in soils.

# Microfloral Distribution within Soil Aggregates

Soil aggregates are convenient units to study the soil microhabitat, because they contain pores of various sizes and individual aggregates can be easily handled. In this section the distribution of microorganisms other than protozoa within aggregates is discussed.

Tyagny-Ryadno (1962) investigated the distribution of bacterial cells in aggregates by immersing dried soil aggregates in collodion so that each aggregate became coated with a thin layer of this material. The treated aggregates were then shaken in water for about 5 minutes. Microbial cells present in the supernatant water were believed to be derived from the outer region of the aggregates. Microbial cells that were retained by the soil aggregates were considered to originate from inside the aggregates. This inovative approach was marred by the fact that the collodion was dissolved originally in a toxic mixture of ethanol and ether. Hattori (1963) proposed an alternative method of determining the distribution of microbial cells within soil aggregates. Soil aggregates were very gently washed in sterile water. Although some slaking occurred, the water-stable microaggregates were not disrupted. After aggregates had sedimented, the supernatant was replaced by sterile water and the gentle dispersion of microbial cells from the aggregates was repeated several times. The numbers of most microorganisms in the supernatant decreased with increasing numbers of washings. After five to ten washes, the remaining microbial cells in the aggregates were dispersed by sonication. The proportion of cells obtained after sonication compared with that obtained from all the washes varied from more than ten to a negligible value depending on the microorganisms involved and the physico-chemical conditions or soil treatments. The washed fractions are believed to consist of cells from the aggregate surfaces and larger pores in the superficial region of aggregates. We shall refer to this part of the aggregate microhabitat as the outer microhabitat (OM), where microbial migration and interchange between these sites and the surrounding environment would be easy. The sonicated aggregate fractions are believed to include cells from the smaller pores and larger pores closed the exterior. We shall refer to this part as the inner microhabitat (IM), in which microbial migration and interchange between these sites and the surrounding environment would be restricted. Our aggregates were easily distintegrated into microaggregates by slaking, when they were immersed in water. Thus the OM and IM in our washing–sonication procedure are concerned mainly with microaggregates. In soils in which water-stable aggregates are not readily formed, this procedure cannot be applied effectively and some other fractionation procedure, such as that described by Hopkins *et al.* (1991a, 1991b), should be used. If the microbial cells are very firmly adsorbed on to aggregate surfaces, most of the population will occur in the sonicated fraction and this is another disadvantage of the washing–sonication procedure. Sonic dispersion may also destroy many organisms, if the sonication energy and exposure time are not limited to as little as necessary to distintegrate the micro-aggregates.

The critical demarcation between smaller and larger pores can be determined roughly from the distribution of introduced cells within sterilized aggregates and the relationship between moisture content and the soil water potential (Hattori and Hattori, 1976). This estimation is based on the fact that the partition ratio between the OM and IM of introduced bacterial cells is constant, if the volume of introduced cell suspension is fixed. From the partition ratio, we can determine the maximum volume of water that has entered the smaller pores. The relationship between the soil water content and the water potential, the water release characteristic, is obtained by the conventional procedure (Russel, 1939). The maximum volume of water that has entered the smaller pores can be related to an equivalent water potential. Although the soil water potential is composed of matric, osmotic, gravitational and other potentials, matric potential can be approximately related to pore diameter by assuming that the values of other potentials and the surface tension of the pore surface are constant, as expressed in the following equation (Hillel, 1971):

$$d = 300/\psi m \qquad\qquad (1)$$

where $d$ is the diameter in μm of the pore-neck of the largest pores filled with soil solution and $\psi m$ is the matric potential (kPa). The estimated value for the demarcation between smaller and larger pores is 3–6 μm in diameter of the pore-neck. Only bacterial cells, some small amoebae and small flagellates would be expected to enter smaller pores than this size.

With the washing and sonication procedure, we have made a series of studies of the distribution and dynamics of bacterial populations in soil aggregates (Hattori, 1973). In experiments with soil aggregates freshly obtained from field soil, two important characteristics were noted. Firstly, the number of bacterial cells in the sonicated fraction was often more than ten times larger than that in the washed fraction. Secondly, most of the bacteria in the sonicated fraction were Gram-negative, whereas most bacterial cells in the washed fraction were Gram-positive. Previously, Gram-positive bacteria have usually been reported as the predominant bacteria in soil. Probably, this was caused by the use of incompletely dispersed soil suspensions in counting methods. This Gram-positive and -negative bacterial distribution can be interpreted in terms of the size of soil pores that harboured the cells in the washed and the sonicated fractions, respectively. Since the cells in the washed fraction are believed to come from the OM and should be liable to desiccation, it would be vital for their survival to possess a resistant form, such as a spore, or to be protected by clay particles or organic macromolecules. By contrast, the cells in the sonicated fraction derived from the IM would be likely to be in a more moist situation and so vegetative non-sporing Gram-negative bacterial cells would be more likely to survive in the IM.

The fungal distribution within soil aggregates was also examined by the washing–sonication technique (Hattori, 1967, 1973). The colony-forming units enumerated by plate counting were probably mostly derived from conidia. The fungi released by the washing treatment were mostly dry spores, but mucilaginous fungal spores, e.g. *Fusarium* spp., were mainly present in the sonicated fraction. It is probable that most fungi producing dry spores reside in the OM and that some of the fungi producing muci-laginous spores, more firmly attached to the soil particles, may reside further inside the aggregates. Since conidiophores are several micrometres in diam-eter, they cannot be accommodated in the smaller pores (<3–6 µm). These results are in harmony with observations of Williams *et al.* (1965), who used the soil washing technique for fungal spores in soil.

## Microscopical Observations of Soil Aggregates

Examination of thin sections of soil aggregates with light or electron micro-scopes can provide more direct information about the distribution of microbial cells in aggregates. Observations are usually limited to a small number of sections and conclusions about soil aggregates in general should be tempered with caution and statistical tests. Kilbertus and his coworkers observed bacterial cells in soil aggregates, especially Gram-negative bacteria mainly in the IM surrounded by mucilaginous slime (Balkhi *et al.*, 1978). Foster (1988) similarly noted that some bacteria were surrounded by ori-ented layers of clay particles embedded in a carbohydrate gel. He observed that whereas microvoids in the rhizosphere contained mixed populations, in the bulk soil similar microvoids usually contained one type of microor-ganism; microfauna and fungi were mainly confined to larger voids. Although neither of these studies involved statistical tests, they did indicate how some bacteria were distributed in the IM. Kilbertus (1980) made a statistical study of the distribution of bacterial cells in thin sections of aggregates from three different soils: a brown acid soil, a rendzina forest soil and a chernozem. He concluded that more than 90% of bacterial cells were present in pores of less than 5 µm in diameter. He also concluded that 80% of the bacterial colonies were present in pores less than 1 µm in diameter. Probably most bacterial cells occur singly or in microcolonies of a few cells that only divide once or twice while they are inside these microniches. This conclusion is consistent with the hypothesis that bacterial cells in general divide once or twice per year in soil (Gray, 1976) and that the time lag required to initiate growth of half of the population with appropriate nutrients may be more than a few weeks for most groups of soil bacteria (Hattori, 1988).

# Distribution and Behaviour of Protozoa within and amongst Soil Aggregates

Detailed studies of microhabitats within soil aggregates play an important first step in understanding protozoan behaviour in the heterogeneous soil system in general. We have been able to determine the distribution pattern of protozoa in soil aggregates and their likely predation of other organisms in the soil.

We shall consider first the distribution and the behaviour of protozoa within soil aggregates and later their capacity to move from one aggregate to another.

## Distribution of protozoa within soil aggregates

It seems likely that protozoa inhabit pores of different sizes in accordance with their own dimensions. It is not easy to confirm this hypothesis in practice, because most protozoan cells in soil are encysted and difficult to identify in that state. By using the washing–sonication technique, we can obtain some indication of how protozoan cells of different sizes are distributed within soil aggregates. Vargas and Hattori (1991) estimated by the most probable number (MPN) method the populations of amoebae, flagellates and ciliates in the washed and sonication fractions. Sonication was limited to 9 KHz (40 watt) for 40–50 seconds to avoid destroying protozoa by this treatment and the viability of *Colpoda* spp. cells inoculated into soil was not affected appreciably. The counts of these protozoan groups in successive washes decreased with increasing numbers of washes (Fig. 3.1). In the case of amoebae and flagellates, the slope was not as steep as in the case of *Colpoda* spp. and other larger ciliates. Appreciable numbers of amoebae or flagellates cells were released from soil aggregates between washings 6 and 9. Such a delayed release probably represents the removal of cells from the IM, although more than 90% of these protozoa resided in the OM. In contrast, cells of *Colpoda* spp. and other larger ciliates were rapidly released and were not detected after the third wash. The rapid release of larger protozoan cells suggests that these cells inhabit the OM. As expected, the washing curve of smaller ciliates showed an intermediate position between large ciliates and amoebae. Darbyshire *et al.* (1989) reported that flagellates in non-disrupted soil aggregates were more resistant to chloroform fumigation than ciliates. This difference may reflect differences between the microhabitats of ciliates and flagellates in soil aggregates.

51

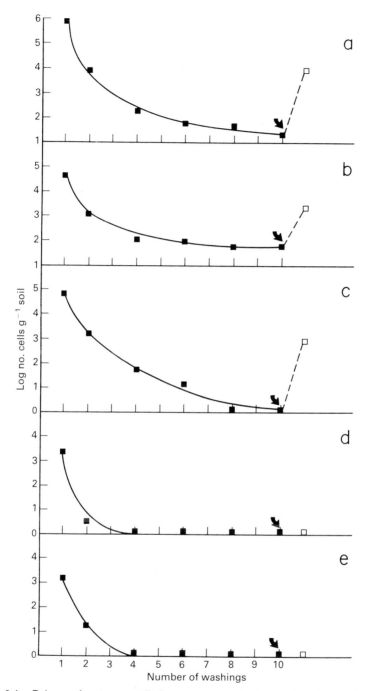

**Fig. 3.1.** Release of protozoan cells from soil aggregates by serial washing and sonication: (a) amoebae; (b) flagellates; (c) small ciliates; (d) *Colpoda* spp.; (e) large ciliates. Arrows mark end of washing treatments and open squares depict populations after sonication. (Vargas and Hattori, 1991.)

**Table 3.1.** The number and frequency of protozoan cells in 1 g dry wt of soil aggregates.[a]

| Protozoan group | Number of cells[b] | Frequency (%)[c] |
|---|---|---|
| Amoebae | 123000 (172) | 67.3 |
| Flagellates | 27300 (38) | 80.6 |
| Small ciliates | 12100 (17) | 42.2 |
| *Colpoda* spp. | 5720 (8) | 45.4 |
| Large ciliates | 1090 (1.5) | 39.6 |

[a] Vargas and Hattori (1990). Protozoan populations estimated by the most probable number method.
[b] Mean number of aggregates in 1 g dry wt was 714. Figures in parentheses are mean numbers of protozoan cells in one aggregate.
[c] Frequency was determined by incubating aggregates separately with the addition of prey bacteria.

In our washing–sonication experiments, aggregates of 1–2 mm diameter were apt to rupture by slaking when immersed in water and the remaining water-stable microaggregates were mostly less than 0.1 mm in diameter. In the inside of such small fragments, there may be a few pores large enough to accommodate large protozoan cells. Further washing–sonication experiments with aggregates of different sizes are obviously required for a more complete understanding of protozoan distribution.

### Distribution of protozoa between soil aggregates

Vargas and Hattori (1990) examined the distribution of different protozoan groups in soil aggregates. As shown in Table 3.1, the amoebae, flagellates, small ciliates, *Colpoda* spp. and large ciliates other than *Colpoda* were present, respectively, in the following frequency percentages: 67.3, 80.6, 42.2, 45.4 and 39.6. These frequencies were much smaller than expected from the mean numbers per aggregate of these groups calculated from MPN counts using 1 g of soil aggregates. These results suggest that the individual cells of these protozoan groups were not randomly distributed and were concentrated in a few aggregates. Nevertheless, the larger protozoa may move very quickly between aggregates and colonize the other aggregates when interconnecting water films are temporarily formed after rain or irrigation.

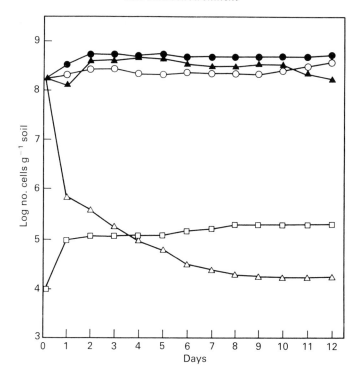

**Fig. 3.2.** Populations of *Klebsiella aerogenes* and *Colpoda* sp. inoculated into sterilized air-dry soil aggregates. Means of four replicates are indicated. Bacterial populations without ciliates in outer region (○) and inner region (•) of aggregates. Bacterial populations in presence of ciliates in outer region (△) and inner region (▲) of aggregates. Ciliate populations (□). (Vargas and Hattori, 1986.)

## *Predation of bacteria by* Colpoda *sp. within soil aggregates*

Protozoan predation of bacteria may be conveniently studied in sterilized air-dried soil aggregates (SADA) for two reasons:

1. Bacterial cells inoculated into SADA colonize both the IM and the OM and the partition ratio is a function of the volume of cell suspension introduced.
2. If sterile water is added before the cell suspension, the number of cells residing in the IM will decrease.

When *Colpoda* sp. cells (32×17×11 μm) were added to soil aggregates, they were present only in the OM and absent from the IM. Fig. 3.2 shows that *Colpoda* sp. devoured bacterial cells only in the OM and cannot feed in the

IM. Since the upper limit of pore-neck diameter of smaller pores is 3–6 μm, *Colpoda* cells were probably too large to penetrate the IM.

In Fig. 3.2 there was a minimal level of surviving bacterial prey. This level was lowered with increases in the number of *Colpoda* present, as shown in Fig. 3.3. Notably, when the average number per aggregate of *Colpoda* cells was 0.5, more than 90% of the prey cells were consumed and this suggests that the predators migrated along water films between aggregates to eat their prey. Why the *Colpoda* were unable to consume all the prey cells in the OM cannot be explained. One possibility is that, if some parts of the aggregate surfaces were hydrophobic, the water films would be interrupted in places and *Colpoda* motility would be restricted (Fig. 3.4). Such local hydrophobic areas due to organic coatings were observed by Bond (1964), who measured contact angles of soil particles. According to Tschapek (1983), water repellancy in soil is mainly caused by amphiphilic humic acid. Humic acids have both hydrophilic (–OH, –COOH and –NH$_2$) and hydrophobic groups.

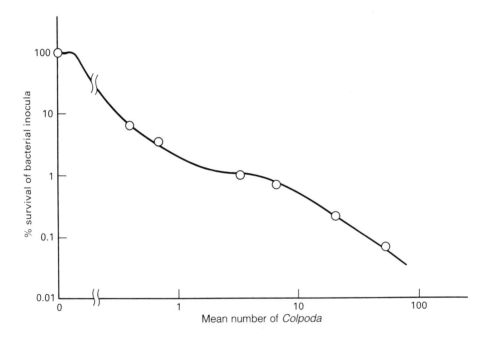

**Fig. 3.3.** Percentage survival of bacteria from original inocula in outer region of soil aggregates in presence of different numbers of *Colpoda* sp. Experimental details as in Fig. 3.2. Both populations on logarithmic scales and estimated after 8 days incubation. Numbers of *Colpoda* sp. expressed as means per aggregate. (Based on data of Vargas and Hattori, 1986.)

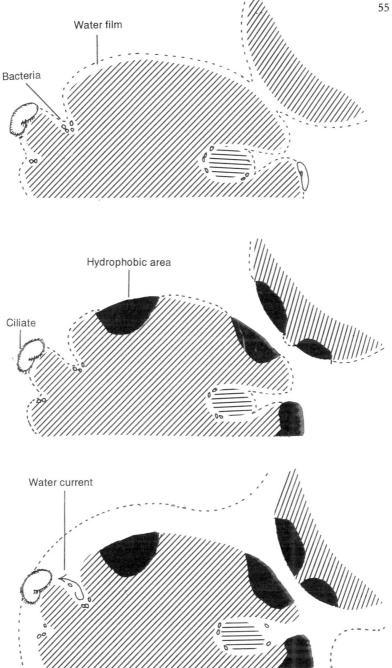

**Fig. 3.4.** Three hypothetical environmental niches for protozoa amongst soil aggregates with water films and hydrophobic regions of different dimensions. Hatched areas represent soil matrix.

An increase in the soil moisture content can greatly affect *Colpoda* predation. This was especially noticeable when the average number of introduced ciliate cells per aggregate was less than 1.0. Since some aggregates have no ciliates in this situation, the predation of bacterial cells in these aggregates would be attributed to migratory protozoa from other aggregates. In a series of experiments summarized in Fig. 3.5, about 48% of aggregates were supposed to contain no *Colpoda*, from the table for a Poisson distribution. With a soil moisture content of 60% water holding capacity (WHC), *Colpoda* sp. devoured about 90% of the bacterial cells in the OM. When the moisture content was increased to 65% WHC, the percentage of consumed bacteria was 99%. These results suggest that the increase in the moisture content increased the size of water films and provided connections between separate aggregates for the *Colpoda*.

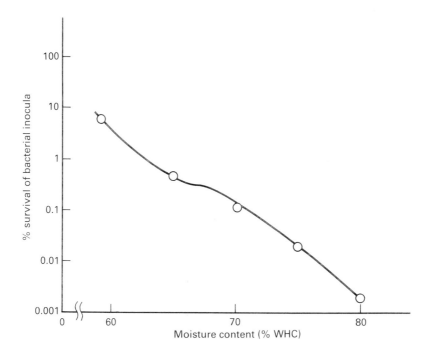

**Fig. 3.5.** Percentage survival of bacterial inocula at different soil moisture contents. Experimental details as in Fig. 3.2. Bacterial populations plotted on a logarithmic scale. Water content expressed as % of maximal water holding capacity (WHC). (Based on data of Vargas and Hattori 1986.)

## Bacterial predation in the OM at sites inaccessible to protozoa

It is suggested that the OM contains pores of various sizes with pore-necks smaller than 3–6 μm in diameter. Protozoan cells cannot enter pores smaller than their usual dimensions unless they are very flexible, such as amoebae. The mean size of *Colpoda* sp. used in our experiments was 32×17×11 μm. It is unlikely that these ciliates can eat bacterial cells inhabiting pores smaller than the predator cells, unless they can generate sufficiently strong ciliary currents in the soil solution to draw bacteria from inaccessible soil pores. Whether ciliates or other soil protozoa can consume bacterial cells while these prey are attached to the surfaces of soil aggregates is unknown. Caron (1987), however, studied the ability of four marine heterotrophic microflagellates (*Bodo* sp., *Cryptobia* sp., *Monas* sp. and *Rhynchomonas nasuta*) to eat the bacterium *Pseudomonas halodurans* when attached to chitin particles or unattached in laboratory cultures. He found *Cryptobia* sp. and *Monas* sp. could easily eat the unattached bacteria, but showed little ability to graze attached or aggregated bacteria. By contrast, *Bodo* sp. and *Rhynchomonas nasuta* showed marked preferences for attached and aggregated bacteria, but only a limited ability to graze unattached bacteria. Similar differences in feeding behaviour between soil protozoa are likely to exist and await investigation.

## Physico-chemical factors affecting protozoan behaviour in soil aggregates

Since most soil protozoa are aerobic, they probably cannot remain active in an anaerobic region of aggregates for an extended period. The volume of the anaerobic region is a function of aggregate size, respiration rate, oxygen concentration at the surface of the aggregate and the oxygen diffusion coefficient. The oxygen diffusion coefficient is affected largely by the soil moisture content and the pore size distribution within the aggregates. Tiedje and his colleages calculated from oxygen measurements and from denitrification rates that appreciable volumes of water-saturated aggregates were anaerobic. Within water-saturated aggregates of 7.0 to 13.0 mm radius, they measured relative anaerobic radii from 1.3 to 5.5 mm for a cultivated silt loam and from 0 to 14.0 mm for an adjacent uncultivated native prairie (Tiedje *et al.*, 1984). The volume of the anaerobic zone and the oxygen concentration at the surface of aggregates will change with moisture content of the soil, as discussed by Smith (1980). If such anaerobic zones are formed within soil aggregates under field conditions, these zones will coincide with part at least of the IM and the lack of oxygen may exclude active protozoan life from this region.

The concentrations of mineral salts or organic substances in the soil solution amongst soil aggregates are likely to show some spatial variation in most soils. Vedy and Bruckert (1982) classified the soil solution into two categories, gravitational and capillary water. They noted differences in the concentrations of organic and mineral components of the soil solution in various soils and at different depths. Menzies and Bell (1988) evaluated two techniques for extracting soil solutions (centrifugation and immiscible liquid displacement) and concluded that the two techniques extract water from different ranges of water potential. The differences between the compositions of the soil solutions obtained by two techniques varied with the soil moisture and any rewetting pretreatment. Such studies of the soil solution strongly suggest that there are two major 'immobile' and 'mobile' categories, where the soil solution percolates very slowly through small pores and more rapidly through larger pores, respectively (Becher, 1985). The conditions for protozoan life are likely to be very different in these two categories of soil solution. The gradients of solute concentration between the two categories may induce some chemotaxic movement of protozoa towards more favourable sites.

## Future Investigations of the Soil Microenvironment

I have tried to consider the existing evidence about the distribution and the behaviour of protozoa in the soil microenvironment as well as for other soil microorganisms. Most of our experiments involved ill-defined factors, but these investigations can be regarded as an initial step in the characterization of soil microenvironments. In this last section, I consider some aspects that clearly need to be explored in the future.

The first aspect relates to the variation in the degree of aggregation amongst soils, effects of the sizes of aggregates and the dynamic nature of soil aggregate formation and destruction. Most soils consist of different sizes of aggregates. In our studies, we used aggregates of 1–2 mm in diameter exclusively. If we had examined larger aggregates, it is possible that different distributions for each protozoan group would have been observed. Other structural units besides aggregates may play important roles in determining protozoan behaviour in soil. We also confined our studies to water-stable aggregates, because the structure of water-labile aggregates was destroyed during the washing treatments. Some soils have few or no water-stable aggregates and it remains to be determined how protozoa behave and are distributed in such soils. Extensive observations of soil thin sections is one way to increase our knowledge of the microenvironments available to soil protozoa. Another way is the use of artificial soil aggregates with a restricted range of pore diameters to observe microbial cells *in situ* within these artificial aggregates, as suggested by Darbyshire *et al.* (1993). It should also be noted that water-stable aggregates themselves are not stable for a long period, but are continually changing. Their structure will be built up by the actions of plant roots or some microorganisms and impaired by wetting and drying, freezing and thawing, or the action of other microorganisms. It is not clear whether soil protozoan action is involved in the mechanisms of aggregate dynamics. The distribution of protozoa in soil aggregates will need to be studied in relation to the degree of aggregation.

The second aspect is concerned with the need for direct investigations of soil microenvironment. Most of our results were obtained by indirect methods. The recognition and the enumeration of protozoa or bacteria in the IM were made on dispersed aggregates. The size of pores where the microbial cells were present was calculated using the water release characteristics. Such assumptions are oversimplifications and are particularly unsuitable for clay soils, which are liable to shrink under suction. Moreover, we cannot consider protozoan movement in detail amongst aggregates without increased knowledge of the size of water films and the state of water in different soil pores. New methods of observing protozoa directly amongst soil aggregates are required. The feeding strategies of different soil protozoa are likely to vary considerably. The connectivity of pores is also an important factor restricting protozoan behaviour within aggregates. Pore connections within soil aggregates seem to be very complex. Darbyshire and his colleagues studied pore connectivities from two-dimensional montages of soil aggregates. In an aggregate of 1 mm$^3$ size, they estimated that 17% of the pore space was large enough to accommodate a protozoan with cross-sectional diameter 20 μm, but only 11% of the pores would be accessible to this protozoan through pore connections with the exterior of the aggregates (Glasbey *et al.*, 1991). Recently, several authors (Bartoli *et al.*, 1991; Young and Crawford, 1991) have demonstrated that the soil-pore network can be categorized by fractal geometry, as introduced by Mandelbrot (1982). We may expect that such quantitative studies of the soil-pore system will promote a better understanding of pore connectivity and tortuosity amongst soil aggregates.

The third aspect is concerned with the utility of the soil aggregate as a microcosm of the life in the soil. A single aggregate often accommodates a range of bacteria, protozoa and fungi. Many of these organisms are involved in either the decompostion, assimilation, oxidation or reduction of various moieties of soil organic matter. Amongst protozoa, bacteria and fungi, we may expect biological interactions, such as predator–prey relationships, symbiosis or parasitism. Microbial migration between aggregates may often be restricted by either soil moisture, other abiotic or biotic factors. Thus, it is possible to imagine that, on occasions, a single aggregate may represent an isolated world or microcosm. If the concept is correct, then more studies of single aggregates by soil microbiologists would be warranted. From recent studies of individual aggregates two interesting facts have emerged (Hattori and Hattori, 1993). One is concerned with the diversity of protozoan fauna in aggregates and the other with the delay in protozoan excystment after immersion of aggregates in water. In experiments where we successively immersed 48 individual aggregates in 1 ml of a mineral solution for several days and then air-dried the aggregates for about one month up to six times, we found that every aggregate contained more than three groups of protozoa and the composition varied in every aggregate except in one instance. Although there is a possibility of not detecting some groups that were strongly attached to soil particles, there is clearly some variation in the distribution of protozoan fauna among aggregates. On the other hand, every aggregate contained more than three groups of bacteria with different growth rates from fast to very slow-growing forms. Another interesting fact that emerged is that the excystment of every protozoan group does not occur immediately after immersion, except for one group of flagellates, but occurs more or less sporadically over 4 to 6 weeks. Vargas and Hattori (1988) showed that excystment of soil amoebae, flagellates, small ciliates and large ciliates followed first order kinetics. This suggests that the probability of excystment in a unit time can be predicted by the respective reaction constant. Active protozoa can encyst again in soil soon after excystment. Fast growing groups of bacteria, in contrast, began to reproduce immediately after immersion and are followed by the slower growing bacterial groups, which may later replace or coexist with the early colonists. Finally, very slow-growing bacterial groups became dominant in most aggregates. Many of the microbial interactions involved in these successions are still incompletely understood.

Soil furnishes a series of transitory habitats consisting of air-filled spaces occupied by microarthropods and larger fauna and some more moist pores that can accommodate a diverse community of microorganisms where organic matter is transformed to provide nutrients for plant growth and the microflora are consumed by soil fauna (Bamforth, 1988). The interactions between soil and microflora as well as fauna are complex. In the analysis of the interactions, the concept of habitable pore space has been widely applied. For example, Elliott *et al.* (1980) noted that nematode growth was increased in soil by the presence of amoebae. They suggested that protozoa provided extra food for nematodes by feeding on bacteria within the pores inaccessible to nematodes.

We may, therefore, conclude that the aggregate structure and the pore network of soil will remain key concepts in future analyses of the soil microhabitat for both micro- and macrofauna.

# Acknowledgement

The author wishes to thank Dr J.F. Darbyshire for his valuable advice on the manuscript for this chapter.

# References

Balkhi, M.E.L., Managenot, F., Proth, J. and Kilbertus, G. (1978) Influence de la percolation d'une solution de saccharose sur la composition qualitative et quantitative de la microflore bactérienne d'une soil. *Soil Science and Plant Nutrition* 24, 15–25.

Bamforth, S.S. (1988) Interactions between protozoa and other organisms. *Agriculture, Ecosystems and Environment* 24, 229–234.

Bartoli, F., Phillippy, R., Doirisse, M., Niquet, S. and Dubuit, M. (1991) Structure and self-similarity in silty and sandy soils: the fractal approach. *Journal of Soil Science* 42, 167–185.

Becher, H.H. (1985) Möglich Auswirukungen einer schnellen Wasserbewegung in Boden mit Makroporen auf den Stofftransport. *Zeitschrift für deutschen Geologischen Gesellschaft* 136, 303–309.

Bond, R.D. (1964) The influence of microflora on the physical properties of soils. *Australian Journal of Soil Research* 2, 123–131.

Caron, D.A. (1987) Grazing of attached bacteria by heterotrophic microflagellates. *Microbial Ecology* 13, 203–218.

Cook, R.J. and Papendick, R.I. (1970) Soil water potential as a factor in the ecology of *Fusarium roseum* f.sp. *cerealis* 'culmorum'. *Plant and Soil* 32, 131–145.

Cutler, D.W. and Crump, L.M. (1935) *Problems in Soil Microbiology.* Longmans Green, London.

Darbyshire, J.F. (1976) Effect of water suctions on the growth in soil of the ciliate *Colpoda steini*, and the bacterium *Azotobacter chroococcum*. *Journal of Soil Science* 27, 369–376.

Darbyshire, J.F., Griffiths, B.S., Davidson, M.S. and McHardy, W.J. (1989) Ciliate distribution amongst soil aggregates. *Revue d'Écologie et de Biologie du Sol* 26, 47–56.

Darbyshire, J.F., Chapman, S.J., Cheshire, M.V., Gauld, J.H., McHardy, W.J., Paterson, E. and Vaughan, D. (1993) Methods for the study of the interrelationships between microorganisms and soil structure. *Geoderma* 56, 3–23.

Duclaux, E. (1898) *Traité de Microbiologie.* Masson, Paris.

Edwards, A.P. and Bremmer, J.M. (1967) Microaggregates in soils. *Journal of Soil Science* 18, 64–73.

Elliott, E.T. and Coleman, D.C. (1988) Let the soil work for us. *Ecological Bulletins* 39, 23–32.

Elliott, E.T., Anderson, R.V., Coleman, D.C. and Cole, C.V. (1980) Habitable pore space and microbial trophic interactions. *Oikos* 35, 327–335.

Emerson, W.W. (1959) The structure of soil crumbs. *Journal of Soil Science* 10, 235–244.

Foster, R.C. (1988) Microenvironments of soil microorganisms. *Biology and Fertility of Soils* 6, 189–203.

Frankel, C. (1887) Untersüchungen über das Vorkommen von Mikroorganismen in verschiedenen Bodenschichten. *Zeitschriften für Hygiene* 2, 521–582.

Glasbey, C.A., Horgan, G.W. and Darbyshire, J.F. (1991) Image analysis and three-dimensional modelling of pores in soil aggregates. *Journal of Soil Science* 42, 479–486.

Gray, T.R.G. (1976) Survival of vegatative microbes in soil. In: Gray, T.R.G. and Postgate, J.R. (eds), *The Survival of Vegetative Microbes.* Cambridge University Press, Cambridge, pp. 327–364.

Greenwood, D.J. (1968) Measurement of microbial metabolism in soil. In: Gray, T.R.G. and Parkinson, D. (eds), *The Ecology of Soil Bacteria.* Liverpool University Press, Liverpool, pp. 138–157.

Griffin, D.M. (1963) Soil moisture and the ecology of soil fungi. *Biological Reviews* 38, 141–166.

Griffin, D.M. (1972) *Ecology of Soil Fungi.* Chapman and Hall, London, pp. 71–112.

Hattori, R. and Hattori, T. (1963) Effect of a liquid–solid surface on the life of microorganisms. *Ecological Reviews* 16, 63–70.

Hattori, R. and Hattori, T. (1993) Soil aggregates as microcosms of bacteria–protozoa biota. *Geoderma* 56, 493–501.

Hattori, T. (1963) Bacterial life in soil. *Nougyou to Engei* 38, 299–304.

Hattori, T. (1967) Microorganisms and soil aggregates as their microhabitat. *Bulletin of Institute of Agricultural Research, Tohoku University* 18, 159–193.

Hattori, T. (1970) Adhesion between cells of *E. coli* and clay particles. *Journal of General and Applied Microbiology* 16, 351–359.

Hattori, T. (1973) *Microbial Life in the Soil.* Dekker, New York.

Hattori, T. (1988) *Viable Count: Quantitative and Environmental Aspects.* Science Tech and Springer Verlag, Madison and Berlin.

Hattori, T. and Furusaka, C. (1960) Chemical activities of *E. coli* adsorbed on a resin. *Journal of Biochemistry* 48, 831–837.

Hattori, T. and Furusaka, C. (1961) Chemical activities of *Azotobacter agile* adsorbed on a resin. *Journal of Biochemistry* 50, 312–315.

Hattori, T. and Hattori, R. (1976) The physical environment in soil microbiology: An attempt to extend principles of microbiology to soil microorganisms. *CRC Critical Reviews in Microbiology* 4, 423–461.

Heijnen, C.E., Postma, J. and Van Veen, J.A. (1990) The significance of artifically formed and originally present protective microniches for the survival of introduced bacteria in soil. *Transactions of the 14th Congress of Soil Science* 3, 88–93.

Hillel, D. (1971) *Soil and Water: Physical Principles and Processes.* Academic Press, New York.

Hopkins, D.W., Macnaughton, S.J. and O'Donnell, A.G. (1991a) A dispersion and differential centrifugation technique for representatively sampling microorganisms from soil. *Soil Biology and Biochemistry* 23, 217–226.

Hopkins, D.W., O'Donnell, A.G. and Macnaughton, S.J. (1991b) Evaluation of a dispersion and elutriation technique for sampling microorganisms from soil. *Soil Biology and Biochemistry* 23, 227–232.

Kilbertus, G. (1980) Étude des microhabitats contenus dans les agrégats du sol leur relation avec la biomass bactérienne et la taille des procaryotes presents. *Revue d'Écologie et de Biologie du Sol* 17, 543–557.

Kuikman, P.J., Van Vuuren, M.M.I. and Van Veen, J.A. (1989) Effects of soil moisture regime on predation by protozoa of bacterial biomass and the release of bacterial nitrogen. *Agriculture, Ecosystems and Environment* 27, 271–279.

Lees, H. and Quastel, J.H. (1946a) Biochemistry of nitrification in soil. I. Kinetic study of, and the effects of poisons on, soil nitrification, as studied by a soil perfusion technique. *Biochemistry Journal* 40, 803–815.

Lees, H. and Quastel, J.H. (1946b) Biochemistry of nitrification in soil. II. The site of soil nitrification. *Biochemistry Journal* 40, 815–823.

McLaren, A.D. (1954) The adsorption and reactions of enzymes and proteins on kaolinite. I. *Journal of Physical Chemistry* 58, 129–137.

McLaren, A.D. (1971) Kinetics of nitrification in soil: Growth of nitrifiers. *Soil Science Society of America Proceedings* 35, 91–95.

McLaren, A.D. and Estermann, E.F. (1957) Influence of pH on activity of chymotrypsin at a solid–liquid interface. *Archives of Biochemistry and Biophysics* 68, 157–170.

McLaren, A.D. and Skujins, J. (1968) The physical environment of microorganisms in soil. In: Gray, T.R.G. and Parkinson, D. (eds), *The Ecology of Soil Bacteria.* Liverpool University Press, Liverpool, pp. 3–24.

Mandelbrot, B.B. (1982) *The Fractal Geometry of Nature.* Freeman, New York.

Marshall, K.C. (1971) Sorptive interaction between soil particles and microorganisms. In: McLaren, A.D. and Skjujins, J. (eds), *Soil Biochemistry 2.* Dekker, New York, pp. 409–445.

Menzies, N.W. and Bell, L.C. (1988) Evaluation of the influence of sample preparation and extraction technique on soil solution composition. *Australian Journal of Soil Research* 26, 451–464.

Oades, J.M. (1984) Soil organic matter and structural stability; mechanisms and implications for management. *Plant and Soil* 76, 319–337.

Postma, J., Hok-A-Hin, C.H. and Van Veen, J.A. (1990) Role of microniches in protecting introduced *Rhizobium leguminosarum* biovar. *trifolii* against competition and predation in soil. *Applied and Environmental Microbiology* 56, 495–502.

Quastel, J.H. and Schofield, J.H. (1951) Biochemistry of nitrification in soil. *Bacteriological Reviews* 15, 1–53.

Robert, M. and Chenu, C. (1992) Interactions between soil minerals and microorganisms. In: Stotzky, G. and Bollag, J. (eds), *Soil Biochemistry 7.* Dekker, New York, pp. 307–404.

Rovira, A.D. and Greacen, E.L. (1957) The effect of aggregate disruption on the activity of microorganisms in the soil. *Journal of Australian Agricultural Research* 8, 659–673.

Russel, M.B. (1939) Soil moisture curve for four Iowa soils. *Soil Science Society of America Proceedings* 4, 51–54.

Smith, K.A. (1980) A model of the extent of anaerobic zones in aggregated soils, and its potential application to estimates of denitrification. *Journal of Soil Science* 31, 263–277.

Stotzky, G. (1986) Influence of soil minerals on metabolic processes, growth, adhesion, and ecology of microbes and viruses. In: Huang, P.M. and Schnitzer, M. (eds), *Interaction of Soil Minerals with Organics and Microbes.* SSSA Special Publication No. 17, Soil Science Society of America, Madison, pp. 305–428.

Tiedje, J.M. Sextone, A.J., Parkin, T.B., Revsbech, N.P. and Shelton, D.R. (1984) Anaerobic processes in soil. *Plant and Soil* 76, 197–212.

Tisdall, J.M. and Oades, J.M. (1982) Organic matter and water-stable aggregations in soils. *Journal of Soil Science* 33, 141–163.

Tschapek, M. (1983) Criteria for determining the hydrophilicity-hydrophobicity of soils. *Zeitschrift für Pflanzenernährung und Bodenkunde* 147, 137–149.

Tyagny-Ryadno, M.G. (1962) Microflora of soil aggregates and plant nutrition. *Izvestia Akademii Nauk SSSR* 2, 242–251.

Vargas, R. and Hattori, T. (1986) Protozoan predation of bacterial cells in soil aggregates. *FEMS Microbiology Ecology* 38, 233–242.

Vargas, R. and Hattori, T. (1988) A simulation approach to the recognition of soil protozoa in the ring method. *Bulletin of Japanese Society of Microbial Ecology* 2, 53–55.

Vargas, R. and Hattori, T. (1990) The distribution of protozoa among soil aggregates. *FEMS Microbiology Ecology* 74, 73–78.

Vargas, R. and Hattori, T. (1991) The distribution of protozoa within soil. *Journal of General and Applied Microbiology* 37, 515–518.

Vedy, J.C. and Bruckert S. (1982) Soil solution: Composition and pedogenic significance. In: Bonneau, M. and Soucheir, B. (eds), *Constituents and Properties of Soils*. Academic Press, London, pp. 184–213.

Wallace, H.R. (1956) Migration of nematodes. *Nature (London)* 177, 287–288.

Wallace, H.R. (1958a) Movement of eelworms. I. The influence of pore size and moisture content of the soil on the migration of larvae of the beet eelworm, *Heterodera schachtii* Schmidt. *Annals of Applied Biology* 46, 74–85.

Wallace, H.R. (1958b) Movement of eelworms. II. A comparative study of the movement in soil of *Heterodera schachtii* Schmidt and *Dytylechus dipsaci* (Kuhn) Filipjev. *Annals of Applied Biology* 46, 86–94.

Williams, S.T., Parkinson, D. and Burges, N.A. (1965) An examination of the soil washing technique by its application to several soils. *Plant and Soil* 22, 167–186.

Young, I.M. and Crawford, J.W. (1991) The fractal structure of soil aggregates; its measurement and interpretation. *Journal of Soil Science* 42, 187–192.

# Soil Nutrient Flow  4

B.S. GRIFFITHS
*Soil–Plant Dynamics Group, Cellular and Environmental Physiology Department, Scottish Crop Research Institute, Invergowrie, Dundee DD2 5DA, UK.*

## Introduction

Protozoa have been less frequently studied in soil than in aquatic environments, probably because of the greater difficulties of observing and growing these microorganisms in such an opaque and heterogeneous medium as soil. There are, however, many similarities between soil and aquatic environments; soil is often conveniently regarded by microbiologists as a porous medium at least partly filled with soil solution. Protozoa can be considered as being active in the aquatic environment within the soil. Accordingly, I have frequently referred to the considerably body of information from marine and freshwater environments to help demonstrate the role of soil protozoa in nutrient flows. Starting from the basic question how do nutrients flow through protozoa, I have discussed the evidence derived from simple and complex laboratory experiments through to large-scale field studies that protozoa are involved in soil nutrient cycling. Subsequently, the importance of protozoa for nutrient flows at discrete sites in soil are considered and finally some promising topics for further research are suggested.

## Direct Effect of Protozoa on Nutrient Flow

Protozoan involvement in nutrient flows stems from their feeding activities; the direct effects of protozoa arise from the fate of ingested food material. In common with other heterotrophic organisms, energy derived from food flows through protozoa according to the 'universal' model of ecological

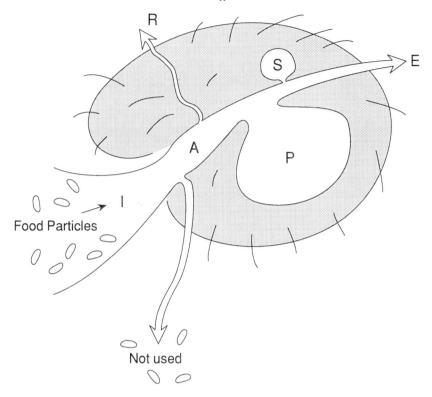

**Fig. 4.1.** The flow of food energy through protozoa. Most of the ingested food (I) is assimilated (A) and used for production (P), respiration (R) and storage (S). Undigested material and nutrients in excess are excreted (E).

energy flow (Odum, 1971; Fig. 4.1). Not all the food ingested by protozoa is utilized; some can be egested apparently unaltered (Heal and Felton, 1970; Caron *et al.*, 1985; Hekman *et al.*, 1992; Zwart and Darbyshire, 1992). These undigested food particles can later become re-available for ingestion by the same or other protozoa. Digested food is used either for cell maintenance and respiration of carbon dioxide, or for the production of new protozoan biomass or storage reserves. Another portion of the food is excreted and usually consists of the structural residues left after digestion (Nilsson, 1987). Generally out of all the carbon ingested, 30% would be respired, 30% excreted and 40% used for production (Sleigh, 1989).

If it is assumed that the protozoan cell has the same elemental ratio of carbon:nitrogen:phosphorus (C:N:P) as its prey, then the protozoan will require sufficient N and P to balance the amount of C used for production. Thus, 40% of the ingested nutrients are usually needed to produce new biomass, to match the 40% of ingested C that goes towards production. This

**Table 4.1.** The effect of C:N ratio in prey and protozoa on protozoan excretion of nitrogen. The assumption is made that the protozoa have consumed prey containing 100 arbitrary units of carbon with an ecological growth efficiency of 40%.

| C:N | | Arbitrary units of N | | | % consumed N that is excreted |
|---|---|---|---|---|---|
| Prey | Protozoa | Consumption | Production | Excretion | |
| 3 | 3 | 33.3 | 13.3 | 20 | 60 |
| 5 | 3 | 20 | 13.3 | 6.7 | 34 |
| 10 | 3 | 10 | 13.3 | N deficient | – |
| 3 | 5 | 33.3 | 8 | 25.3 | 76 |
| 5 | 5 | 20 | 8 | 12 | 60 |
| 10 | 5 | 10 | 8 | 2 | 20 |
| 3 | 10 | 33.3 | 4 | 29.3 | 88 |
| 5 | 10 | 20 | 4 | 16 | 80 |
| 10 | 10 | 10 | 4 | 6 | 60 |

leaves 60% of the ingested nutrients, surplus to protozoan requirements, to be excreted. This is the basis of direct nutrient flow through protozoa at either the cell, population or community level. The proportion of nutrient in surplus is affected by the relative nutrient status of the prey and the protozoa. If the protozoa has a higher C:nutrient ratio than the prey, then more N and P would be in excess and more than 60% of the ingested nutrient will be excreted. When the ratio of C:nutrient increases equally in both protozoa and prey, the amount of nutrient excreted decreases even though it is still 60% of that ingested. Some examples of the possibilities are given in Table 4.1 with regard to carbon and nitrogen. There is experimental evidence indicating that the C:nutrient ratio of the prey affects the direct contribution of protozoa to nutrient flows, as would be predicted from theoretical considerations. Nutrient remineralization into the medium was reduced when the marine flagellate *Paraphysomonas imperforata* was fed with N or P-limited prey (Caron and Goldman, 1988). Substantial amounts of N, but not P, were mineralized during the stationary growth phase of the flagellate, probably reflecting different biochemical roles for N and P. The authors suggested that the flagellate strove towards producing biomass with a specific elemental stoichiometry regardless of the C:N:P of the prey. The results of experiments by Darbyshire *et al.* (personal communication), with the soil ciliate *Colpoda steinii* and bacterium *Arthrobacter* sp. in liquid culture, also show that prey C:N:P influences protozoan excretion. With N-limited bacteria (4.1% N) there was a lag of 2 days before N was mineralized and the final amounts of N released were half that from N-sufficient bacteria (8.5% N). Similar comparisons made in batch and chemostat culture also showed twice as much P mineralization by *C. steinii* from P-sufficient (1.72% P) than from P-limited (0.46% P) bacteria.

The C:N:P ratios of organisms isolated from natural habitats are often variable. The choice of values for soil organisms is made more difficult, because microbial cells cannot yet be sufficiently separated from other soil materials for reliable elemental analyses (Bakken, 1985). A natural assemblage of marine bacteria had a C:N of 3.7 (Lee and Fuhrman, 1987), while a laboratory culture of mixed marine bacteria gave different ratios depending on which nutrient was limiting. Thus, the C:N and C:P ratios were for C-limited bacteria 4.8 and 7.7, for N-limited bacteria 6.7 and 20 and for P-limited bacteria 6.3 and 55, respectively (Bratbak, 1985). Soil organisms have been grown in culture and then analysed, although there is always the uncertainty of whether they would have the same composition when growing in the soil. A range of eight bacteria had a C:N ratio of 3.7 (Bakken, 1985). Van Veen and Paul (1979) found that the C:N ratio of a bacterium varied from 3.2 to 5.2 and that of a fungus from 9.5 to 38, but their respective C:P ratios were 19–43 and 97–293 under different growth and matric potential conditions. The C:N ratios of protozoa in laboratory cultures have also been measured and were similar to the values for bacteria. Band (1959) gave a value of 4.7 for the amoeba, *Hartmanella rhysoides*, while more recent figures of 3.3 for a flagellate, *Bodo saltans*, 3.5–5.2 for three ciliates, *Colpoda steinii*, *C. cucullus* and *Cyclidium glaucoma*, and 3.5 for an amoeba, *Naegleria gruberi*, have been reported (Borkott, 1989). Protozoan C:N has been considered either to be the same as the bacterial prey in some studies, a C:N of 5 by De Ruiter *et al.* (1993) and a N content of 5% (approximately equivalent to C:N 10) by Clarholm (1985), or slightly higher such as a protozoan C:N of 7 and a bacterial C:N of 4.5 (Rutherford and Juma, 1992b). Given the current discrepancies between estimates for soil isolates, a C:N ratio of 5 for both bacteria and protozoa would appear to be the most realistic assumption at present.

It is interesting to note that the C:N ratios quoted earlier are more stable, even under N limitation, than the C:P ratios. This makes any assumptions about the excretion of excess P by protozoa particularly difficult. Another uncertainty is concerned with fungal prey, which can have very high overall C:N ratios due to the presence of empty hyphae. It would seem reasonable to assume that mycophagous protozoa would prefer to consume the cytoplasm-rich hyphal tips and spores rather than other parts of the mycelia. If this is correct, then the actual C:nutrient ratio of the ingested food would be lower than that from the whole mycelia. A low figure of 10 for the C:N ratio would seem to be realistic for fungal prey and even if the C:N ratio of the protozoa is 5, nitrogen would still be excreted (Table 4.1).

All this protozoan consumption and excretion occurs during their short periods of growth, but the recycling of their nutrients after death must also be significant (Schönborn, 1992). Populations of protozoa in soil do fluctuate quite markedly (Clarholm, 1981, 1989; Christensen *et al.*, 1992) and this implies a considerable death rate as well as a large production rate. Dead protozoa would be rapidly decomposed and with their low C:N ratio would result in a net mineralization of N (Bartholomew, 1965). This recycling of N would be over a longer term than direct excretion of excess N. It would also necessarily involve a smaller proportion of the N ingested by the protozoa. The form of the nutrient that is excreted is also important, as this affects its utilization by either microorganisms (bacteria or fungi) or plants. Nitrogen is excreted by protozoa mainly as ammonium (Doyle and Harding, 1937; Stout, 1980), and phosphorus is also considered to be excreted mainly in an inorganic form (Cole *et al.*, 1978). So both these nutrients would be readily available for use by other soil organisms.

## Indirect Effects of Protozoa and a Comparison with Direct Effects

There are two mechanisms by which soil protozoa can indirectly affect nutrient flows. Firstly, because the protozoa are actively removing certain of the microorganisms present, the composition of the microbial community is altered. Secondly, the presence of protozoa may enhance the activity of the remaining microorganisms. Protozoa are known to be selective in the species of bacteria that they ingest and this has been shown for flagellates (Caron, 1987), ciliates (Taylor and Berger, 1976), amoebae (Singh, 1941, 1942) and for natural assemblages of marine protozoa (Sherr *et al.*, 1992). Selection is made on the basis of cell size, usually the larger and dividing cells are consumed (Sherr *et al.*, 1992), cell location, whether attached or in suspension (Caron, 1987), colony morphology, whether single or aggregated cells (Sibbald and Albright, 1988) and also on the chemical composition of the cell, such as the content of pigments (Singh, 1941) or toxins (Curds and Vandyke, 1966). Selective grazing promotes the occurrence of certain species and cell morphologies within the microbial community (Güde, 1979; Bianchi, 1989; Shikano *et al.*, 1990). In general, selective grazing favours small, slow-growing cells and acts to maintain the taxonomic and metabolic diversity of the microflora (Sherr *et al.*, 1992), while allowing different competing bacteria to coexist (Alexander, 1977). The presence of a wider range of microbial species will allow a wider variety of biochemical reactions to occur under different conditions and encourage nutrient mineralization. Besides the importance of a diverse decomposer microflora, a diverse protozoan community is thought to be important in maintaining decomposition rates (Stout, 1980). Protozoan community structure varies, for example, in the rhizospheres of different plant species (Darbyshire and Greaves, 1967;

Gel'tser, 1991). It is uncertain how much this difference in protozoan fauna is induced by the different plant species and serves to control the subsequent microfloral community. Further investigation of this topic may increase our understanding of protozoan interactions in the rhizosphere.

It has been assumed that grazing maintains a bacterial population in a 'youthful' state (Cutler and Bal, 1926; Johannes, 1965) by preventing the accumulation of senescent cells, by keeping the bacteria in the log growth phase and by keeping bacterial abundance low enough to minimize substrate limitation (Sieburth and Davis, 1982). Implicit in this concept is the idea that the remaining 'young' cells are more active than ungrazed cells and evidence is emerging to support this view. Many studies have demonstrated an increase in $CO_2$ evolution in the presence of protozoa (e.g. Telegedy-Kovats, 1932; Singh, 1964; Coleman et al., 1977; Kuikman et al., 1990a) and this is generally regarded as an increase in C mineralization resulting from enhanced bacterial activity. The overall picture, however, may sometimes not be as straightforward. Although Coleman et al. (1978) observed more $CO_2$-C evolution from soil microcosms containing bacteria (Pseudomonas cepacia) plus amoebae (Acanthamoeba polyphaga) than from bacteria alone, the decrease in bacterial biomass caused by amoebal grazing almost exactly balanced the increase in respiration. This could be interpreted as a reallo-cation of C rather than increased mineralization (Table 4.2). Studies of the effects of protozoa on purely bacterial processes demonstrate more conclusively that the activity of individual bacteria is enhanced in the pres-ence of protozoa. As nitrogen fixation by the bacterium Azotobacter chroo-coccum was stimulated in the presence of protozoa (Nasir, 1923), it was proposed that protozoan grazing extended the period of activity of the bacterial cells (Cutler and Bal, 1926). Other mechanisms have been postu-lated such as a reduction in oxygen concentration, modification of pH by excreted ammonia or the release of specific stimulatory compounds (Darbyshire, 1972). Nitrification was stimulated in both a mixed culture of soil bacteria (Griffiths, 1989a; Fig 4.2) and in a gnotobiotic culture of nitrifiers (Verhagen and Laanbroek, 1992) by the presence of protozoa. The mean specific rates of nitrification in soil percolation columns were 3.48

**Table 4.2.** Carbon balance ($\mu$g C g$^{-1}$ soil) in soil microcosms inoculated with bacteria, amoebae and/or nematodes over 24 days (Coleman et al., 1978).

| Biomass C in | | | | |
|---|---|---|---|---|
| Bacteria | Amoebae | Nematodes | Respired $CO_2$ C | Total C |
| 1131 | | | 694 | 1825 |
| 570 | 223 | | 926 | 1719 |
| 702 | | 18 | 1044 | 1764 |
| 283 | 40 | 26 | 1155 | 1504 |

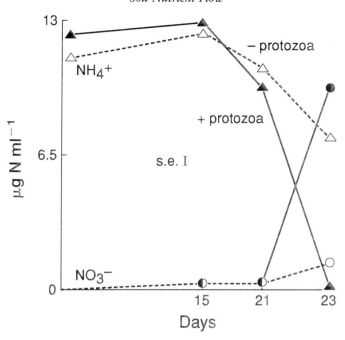

**Fig. 4.2.** Concentration of ammonium (triangles) and nitrate (circles) in an ammonium oxidizer medium inoculated with a mixed culture of bacteria, with (filled symbols) and without (open symbols) protozoa. (From Griffiths, 1989a.)

and 0.72 pMol $NH_4^+$ bacterium$^{-1}$ hour$^{-1}$ with and without the flagellate *Adriamonas peritocrescens*, respectively (Verhagen, 1992). Again the actual mechanism is uncertain. The production of siderophores by the bacterium *Pseudomonas putida* was increased by the presence of the amoeba *Acanthamoeba castellanii* in soil (Levrat *et al.*, 1989) and further work in liquid culture demonstrated that the addition of amoebal culture filtrates to bacterial cultures significantly enhanced siderophore production, ammonification and respiration (Levrat *et al.*, 1992). These authors concluded that the enhancement of bacterial metabolism was due to some unidentified, stimulatory factors from the amoebae. Bacterial productivity is also increased by grazing, as shown by Pussard and Rouelle (1986) for soil protozoa and by Riemann *et al.* (1986) for marine protozoa.

The overall effects of protozoa on nutrient flow have been reviewed many times and can be summarized as:

• To regulate and modify the size and character of the bacterial community (Alexander, 1977; Stout, 1980; Frey *et al.*, 1985; Lousier and Bamforth, 1988; Gel'tser, 1991; Sherr *et al.*, 1992).
• To accelerate the turnover of microbial biomass/soil organic matter/

nutrients (Stout and Heal, 1967; Alexander, 1977; Stout, 1980; Frey
*et al.*, 1985; Porter *et al.*, 1985; Lousier and Bamforth, 1988; Sleigh,
1989; Kuikman, 1990; Vargas, 1990; Zwart and Brussaard, 1991).

* The direct excretion of nutrients (Clarholm, 1983; Beaver and Crisman,
  1989; Kuikman, 1990; Vargas, 1990; Zwart and Brussaard, 1991).
* The re-inoculation of new substrates by the phoretic transport of
  excretion of viable bacteria (Vargas, 1990; Verhagen, 1992).

This latter point, inoculation of new substrates, is a function usually
attributed to larger fauna (Visser, 1985). Recent evidence suggests that protozoa do not move more than a few millimetres in soil (Kuikman *et al.*, 1990b;
Griffiths and Caul, 1993) and may be confined to a particular soil aggregate
(Hattori, 1988). The importance of protozoa in the transportation of bacteria
to new substrates is uncertain at present.

The role of protozoa in the regeneration of phosphorus in mixed cultures
of bacteria and protozoa has been interpreted in several ways. It has proved
difficult in practice to differentiate between the direct and indirect effects
of protozoa. Johannes (1965) discounted the indirect effects of grazing by
the ciliate *Euplotes vannus* on the bacteria and demonstrated that regeneration could be accounted for by the direct excretion of P by protozoa.
Barsdate *et al.* (1974), however, reported that excretion of P by the ciliate
*Tetrahymena pyriformis* was minor and that enhanced mineralization was
due to a higher turnover of P by bacteria in grazed systems. Fenchel and
Harrison (1976) also favoured an indirect effect, suggesting that protozoa
might reduce a limiting factor, excrete a growth-promoting substance or
select for faster growing bacterial species. Cole *et al.* (1978) studied P cycling
in a model soil system and concluded that the direct excretion of inorganic
P by the amoeba *Acanthamoeba polyphaga* was the major pathway, rather
than increased bacterial turnover. Taylor (1986) found that the effect of the
ciliate *Colpidium colpoda* was due to the reduction in bacterial numbers
and that their excretion of P was probably a significant part of the uptake
by bacteria. In a chemostat system, a natural assemblage of flagellates greatly
increased the remineralization of P from dissolved organic matter via the
direct consumption of bacteria (Bloem *et al.*, 1988). Beaver and Crisman
(1989) also concluded that the excretion of organic P by ciliates may represent a substantial pathway for P to be assimilated by phytoplankton.

At present, these conflicting experimental results can best be interpreted
by assuming that a combination of both direct and indirect protozoan
effects occur concurrently in most situations, as depicted in Fig. 4.3. It
remains to be determined under what conditions the direct and indirect
effects predominate.

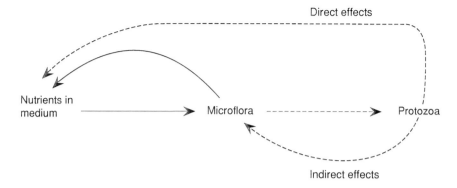

**Fig. 4.3.** Schematic diagram of the nutrient flow between medium, microflora and protozoa. The dashed lines indicate effects of protozoa.

## Laboratory Studies

Increases in the accumulation of nutrients in mixed cultures of bacteria plus protozoa, compared with pure bacterial cultures, have commonly been used to demonstrate protozoan effects on net mineralization. Experiments in liquid batch cultures have shown that the accumulation of nitrogen and phosphorus is enhanced by flagellates (Lawrie, 1935 [nitrogen]; Johannes, 1965 [phosphorus]; Zwart and Darbyshire, 1992 [N]), ciliates (Meiklejohn, 1932 [N]; Doyle and Harding, 1937 [N]; Johannes, 1968 [P]; Griffiths, 1986 [N]; Taylor, 1986 [P]) and amoebae (Levrat, 1990 [N]). In continuous flow chemostat cultures, mixed protozoan populations have also increased nutrient accumulation (Güde, 1985 [N+P]; Bloem *et al.*, 1988 [N+P]. Isotopic labelling has further shown that the turnover of phosphorus, rather than just accumulation, is enhanced by protozoa (Barsdate *et al.*, 1974; Taylor, 1986). Studies of nitrogen turnover are also technically possible, using the stable isotope $^{15}N$, and would be useful for comparing the behaviour of N and P.

Studies have also been made using sterile soil inoculated with bacteria and with or without protozoa. These have again demonstrated increased net mineralization in the presence of ciliates (Griffiths, 1986 [N, but not P]), amoebae (Meiklejohn, 1930 [N]; Coleman *et al.*, 1977 [N+P]; Cole *et al.*, 1978 [P]; Woods *et al.*,1982 [N]; Rutherford and Juma, 1992a [N]) and mixed protozoan populations (Frey *et al.*, 1985 [N+P]; Kuikman and Van Veen, 1989 [N]). Verhagen (1992) found no enhanced mineralization of N following the inoculation of flagellates into soil already containing bacteria, although the soil was only sampled after 14 weeks. Any possible mineralized N may have been immobilized by that time.

Laboratory studies have also proved useful in determining how proto-zoan-induced nutrient flows are affected by different environmental factors. The amounts of available nutrients (N or P) and carbon alter the pattern of nutrient mineralization. In soil microcosms prepared with or without additional glucose C, which presumably led to N-limited or C-limited conditions respectively, grazing by *Acanthamoeba polyphaga* increased the concentration of inorganic-P in the soil after 17 and 24 days incubation, compared with controls containing only the bacterium *Pseudomonas cepacia*. The amount of P mineralized by amoebae was only slightly lower with glucose amendment, compared with the non-amended treatment (Cole *et al.*, 1978). Net N mineralization, however, was so reduced under C-excess/N-limited conditions that $NH_4$–N could only be detected in the bacteria+amoeba treatments, but was not significantly different from the bacteria-only controls after 24 days (Woods *et al.*, 1982). Using the earlier assumptions, that 60% of ingested N is excreted by protozoa with a C:N of 5, Woods *et al.* (1982) calculated that amoebae would release 36 µg N $g^{-1}$ soil under the C-limited conditions. This compared well with the value of 32 µg obtained from their experiments and the difference may be due to bacterial uptake of some of the mineralized N. Making similar calculations for the C-excess/N-limited situation, the data from Coleman *et al.* (1978) and Woods *et al.* (1982) indicate that 65 µg N $g^{-1}$ would have been mineral-ized, most of which would be rapidly assimilated by the bacteria under these conditions in order to account for the observed value of 5 µg. Sinclair *et al.* (1981) made the same observation using similar experimental con-ditions. Nitrogen was mineralized from the bacterium *Pseudomonas pauci-mobilis* under C-limiting conditions when the available C was exhausted (47 hours when grazed by *Acanthamoeba polyphaga* and 71 hours with bacteria only). When N was limiting there was no accumulation of inorganic N and N mineralization could only be deduced from the growth of amoe-bae. The N mineralized by protozoan grazing, because it was in short supply, appeared to be re-utilized by the bacteria in a continuous internal cycle. A mathematical model of nutrient dynamics developed by Hunt *et al.* (1977) showed that predation would increase the accumulation of N, but would have little effect on bacterial growth when C was limited. Pre-dation was also expected to stimulate bacterial growth by releasing N when N was limiting, although there would be no accumulation of N in the medium under these conditions. Their experimental data conformed to the behaviour of this mathematical model. Caron *et al.* (1988) tested similar assumptions in a model aquatic system containing phytoplankton, mixed bacteria and flagellates (*Monas* sp.). The phytoplankton can be thought of as a means of sampling the available N in the system. When N was not limiting, protozoa made no difference to the amount of N in the phyto-plankton. When N was limiting, however, there was significantly more N in the phytoplankton when protozoa were present. They concluded that protozoa would be more important in mineralizing N under N-limited conditions. The same may not be true for P as Bloem *et al.* (1988) measured considerable phosphorus remineralization by flagellates even though P was apparently not limiting.

Protozoa act as prey as well as predators in the soil food web. They are consumed mainly by nematodes and microarthropods, but also compete with these and other microbial-feeding fauna for food. These trophic interactions can affect the flows of nutrients. Elliott *et al.* (1980) stated that the feeding of organisms from one trophic level on another in soil probably accounts for the transfer of much C, N and P in terrestrial ecosystems. These authors used microcosms of coarse- or fine-textured soil inoculated with bacteria, bacterial-feeding amoebae and bacterial/amoebal-feeding nematodes. More nematodes resulted when amoebae were present as food and more nematodes developed in the coarse-textured soil. They suggested that amoebae are able to feed on bacteria in pores inaccessible to nematodes and then emerge to act as food for the nematodes. Thus, soil protozoa provide a link in the food chain akin to aquatic protozoa (Fenchel, 1982; Sleigh, 1989). Nutrient mineralization varies with the complexity of the food web. The addition of nematodes to microcosms containing both bacteria and amoebae reduced net mineralization under C-limited conditions (Woods *et al.*, 1982). They observed an accumulation of 32 µg $NH_4$–N $g^{-1}$ soil from bacteria+amoebae and 16 µg from bacteria+amoebae+nematodes. However, it was calculated that, whereas bacteria+amoebae should have mineralized 36 µg, bacteria+amoebae+nematodes should have released 57 µg. The discrepancy between expected and observed when nematodes were added was assumed to be because of changes in the nematode excretory products. When the nematodes had exhausted their bacterial food supply, in order to conserve C they would excrete nitrogen as $NH_4$–N rather than as amino acid N. As amino acid N is readily available to soil microorganisms, it would appear that N turnover was increased by the presence of nematodes. By contrast, phosphorus was retained by the nematodes and so the turnover of P was not increased (Cole *et al.*, 1978). Protozoa would appear to have a distinct advantage over multicellular bacterial-feeding fauna with their rapid multiplication (generation time, 1–2 days). Nematodes generally take 4–7 days to develop from egg to adult and with only a single bacterial species available for food, protozoa will outcompete nematodes under these conditions (Sohlenius, 1968; Griffiths, 1986). Although numbers of nematodes are reduced by the presence of protozoa, there is a stimulatory effect of nematodes on protozoa (Table 4.3) and this is thought to be due to an indirect effect of nematodes stimulating bacterial production. The accumulation of $NH_4$ from bacteria+ciliates+nematodes was also initially faster than from bacteria+ciliates (Griffiths, 1986). These results differ from those of Woods *et al.* (1982) referred to above, because in this study the protozoa and nematodes were competing at the same trophic level rather than acting

**Table 4.3.** Numbers of bacteria (B), ciliates (C) and nematodes (N) incubated for 8 days in soil extract liquid culture. Mean (± SD) (Griffiths, 1986).

| Treatment | Bacteria (B) $(\times 10^7 \text{ ml}^{-1})$ | Ciliates (C) $(\text{ml}^{-1})$ | Nematodes (N) $(\text{ml}^{-1})$ |
|---|---|---|---|
| B | 17.0 (0.93) | – | – |
| B+N | 24.7 (2.62)* | – | 213 (18)*** |
| B+C | 8.0 (1.11)* | 1950 (409)** | – |
| B+N+C | 16.5 (4.21) | 3700 (492) | 7　(3) |

*, **, ***; significantly different from B+N+C at $P \leqslant 0.05$, 0.01 and 0.001, respectively.

as a food chain. In soil environments, protozoa and nematodes exist together (Christensen *et al.*, 1992; Griffiths *et al.*, 1993) and nematodes can even predominate in the rhizosphere (Griffiths, 1990). This coexistence probably occurs because of their different diets and different microhabitats in the soil (e.g. in different pores and substrates). Consequently, many complex interactions can exist between such organisms of the same trophic level. A more comprehensive study of decomposer food webs in soil involved the inoculation of soil microcosms with bacteria, bacterial-feeding amoebae, flagellates, nematodes and mites (Brussaard *et al.*, 1991). The mites stimulated amoebal numbers due to the increased availability of bacteria and there was a decrease in nematodes in the presence of large numbers of amoebae, as noted by Sohlenius (1968) and Griffiths (1986). Brussaard *et al.* (1991) concluded that the amoebae fed on flagellates and that the presence of mites increased N mineralization.

The structure of the soil in which the protozoa live can affect them in two ways. Firstly, many of the pores are too small for protozoa to enter and this led Postma and Van Veen (1990) to propose the concept of habitable pore space. Secondly, even when the pore is large enough for protozoa a water film has to be present for the organism to be active. The experiments by Elliott *et al.* (1980), described earlier, recognized the importance of soil structure and demonstrated greater flows of C and therefore other nutrients in the coarse-textured soil. The effects of soil texture on N mineralization were specifically tested by Rutherford and Juma (1992a), who used a bacterial inoculum of $^{15}$N-labelled *Pseudomonas* sp. Grazing of bacterial N by *Acanthamoeba* sp. and subsequent mineralization was more effective in coarse than in fine-textured soil. It has been shown that small pores (<6 μm diameter) can act as protective microhabitats against protozoan predation for bacteria introduced into soil (Heijnen and Van Veen, 1991). Also, the degree of C turnover is related to the amount of access that protozoan grazers have to bacteria (Killham *et al.*, 1993). Studies in our laboratory altered the structure of a single soil type rather than compared different soils and have found that only relatively large changes to soil structure, such as that produced by grinding and severe compaction, affect protozoan movement and multiplication (Griffiths and Young, unpublished).

Protozoa require a certain moisture content to be active, related to the size of the water-filled pores (Bamforth, 1985). Amoebae excyst in a matter of hours and grow rapidly to take advantage of favourable moisture conditions (Bryant *et al.*, 1982). Soil pores would, therefore, only have to contain soil solution for a short period of time for trophozoite populations to develop. Kuikman *et al.* (1989) imposed fluctuating moisture conditions by using plants in soil microcosms and varying the length of time between watering. They showed that protozoa were equally effective at mineralizing bacterial N under fluctuating wet and dry or continuously wet conditions. In contrast, the mineralization of N was severely hampered under fluctuating moisture in treatments inoculated with only bacteria. Protozoa rapidly recovered after a drought period and stimulated the turnover of microbial N. In the absence of protozoa, most of the N was immobilized in the bacterial biomass (Kuikman *et al.*, 1991).

Several laboratory studies have shown that N released by protozoa is subsequently taken up by plants, but this has not yet been demonstrated for P in soil. Elliott *et al.* (1979) demonstrated that plants grown in the presence of amoebae obtained more N than plants grown in their absence. This proved that the extra mineralization of N in soil microcosms (e.g. Coleman *et al.*, 1977) was available for plant uptake. Clarholm (1985) also showed that wheat contained more N and had a bigger dry weight in microcosms containing mixed protozoan populations than in those containing only bacteria. It was also shown that bacteria were able to mineralize N from soil organic matter when supplied with sufficient energy, such as from root-exudate. Protozoan grazing seemed to be required to make bacterial N available for plants. Kuikman and coworkers used $^{15}$N-labelled bacteria to investigate the effects of protozoa on the release of bacterial N and soil N. Protozoan grazing increased the quantity of bacterial N available to plants, both N directly released from bacterial cells and N mineralized from soil organic matter as a result of the higher N turnover (Kuikman *et al.*, 1989). Protozoa were more important at increasing plant N under fluctuating than constant moisture conditions, although there was no difference in plant biomass with or without protozoa (Kuikman *et al.*, 1989, 1991). Competition between bacterial-feeding protozoa and nematodes, although affecting the size of their populations, had little effect on the amount of N available to plants (Table 4.4, Griffiths, 1989b). These studies show that protozoa release N from microbial biomass and that this N is later available to plants. Other studies have shown that protozoa can release 30–50% of the N immobilized in the microbial biomass for plant assimilation (Clarholm, 1985; Ritz and Griffiths, 1987; Kuikman *et al.*, 1989).

**Table 4.4.** Nitrogen content of ryegrass (*Lolium perenne*), the abundance of bacteria and the biomass of ciliates and nematodes in soil microcosms after 28 days (Griffiths, 1989b).

| Treatment | Plant N (mg) | Bacteria (B) ($\log_{10}$ g$^{-1}$) | Ciliates (C) ($\mu$g g$^{-1}$) | Nematodes (N) ($\mu$g g$^{-1}$) |
|---|---|---|---|---|
| B | 4.61 | 8.80 | – | – |
| B+N | 5.27* | 8.60 | – | 34.6 |
| B+C | 5.24* | 8.51* | 102.1 | – |
| B+N+C | 5.24* | 8.46* | 122.8 | 4.4** |
| SE | 0.17 | 0.07 | 1.2 | 3.65 |

  * significantly different from B at $P \leqslant 0.05$.
 ** significantly different from B+N at $P \leqslant 0.001$.

# Field Studies and Food Web Analyses

Field studies have adopted two strategies to determine the role of protozoa in nutrient flows. Protozoan effects were inferred from the concomitant changes in protozoa, microflora and nutrients observed over time. Alternatively, certain groups of organisms were removed using selective biocides and the consequent effects on other organisms and nutrients were determined.

Clarholm (1981) reported that 2 days after heavy rain on a podzol planted with Scots pine there was a ten-fold increase in bacterial biomass in the humus layer. There was a 20-fold increase in amoebae 4 days after the rain that coincided with a decrease in bacteria. Amoebae were considered to be important in releasing bacterial N. A similar pattern of events occurred in a field of barley, with a peak of bacterial numbers 2 days after rain and then a subsequent peak of amoebae. In this study, the increase in amoebae was linked to a measured increase in N uptake by the barley plants (Clarholm, 1989). Similar microbial populations have been monitored in semiarid grasslands by Elliott *et al.* (1984) and by Ingham *et al.* (1986a). In both of these studies, an increase in protozoa was accompanied by a decrease in bacterial numbers and an increase in inorganic N to give further confirmation that protozoa can be responsible for the remineralization of bacterial-N under field conditions.

Biocides have been successfully used to remove nematodes and microarthropods from soil, but not protozoa. In a review of the effects of 12 biocides on soil organisms, Ingham (1985) found no evidence for an effective, selective protozoan biocide. Protozoan populations were, however, reduced by a fungicide (benlate) and all the chemicals used had effects on more than one group of organism. Further studies of these biocides in a laboratory microcosm experiment (Ingham and Coleman, 1984) suggested that certain biocides would be useful for some soil organisms, e.g. cyclohex-imide and amphotericin B (fungizone) against fungi, streptomycin against bacteria, cygon against microarthropods and carbofuran against nematodes. Again, no specific protozoan biocide was found. Protozoa have been effec-tively removed from soils with chloroform, although this also affects the microflora (Ingham and Horton, 1987) and probably limits the use of $CHCl_3$ to laboratory studies, for example Darbyshire *et al.* (1989). In a laboratory experiment using soil from a lodgepole pine forest, biocides have been used to show that bacteria accumulated N in their biomass and that a reduction in the size of the bacterial population increased amounts of inorganic N in the soil (Ingham *et al.*, 1986b). Peaks of protozoan populations occurred at the same time as bacteria rather than being delayed by one or two days, as described earlier in the field studies. Fungi were considered to be the major mineralizers of N in this investigation. Biocides have been applied to plots in field situations, but these have not satisfactorally demonstrated protozoan effects. Results from a semiarid grassland demonstrated that bacteria immobi-lized N in their biomass, but it could only be assumed that this was released by faunal grazing (Ingham *et al.*, 1986c). When furadon (fungicide) and cygon (nematicide) were applied to plots in shortgrass prairie, mountain meadow and lodgepole pine forest, different degrees of bacterial dominance occurred in these ecosystems (Ingham *et al.*, 1989). In a no-tillage agroecosystem treated with streptomycin (bacteriocide) and captan (fungicide), the only effect on protozoa was an increase of flagellates in the litter layer (Ingham *et al.*, 1991). Although protozoa were enumerated in these latter two studies, the absence of a specific protozoan biocide and side-effects on other soil microorganisms resulted in little information about the role of protozoa.

Beare *et al.* (1992) used biocides and food web analysis in a study of conventional-tillage and no-tillage systems. The results from control plots, where no biocides were applied, showed that protozoa were the major consumers of bacteria and mineralized most of the bacterial N from buried crop residues. There was a shift to fungal dominance in surface applied residues, which was confirmed by the use of a bacteriocide and a fungicide. Again the importance of protozoa was not revealed by the use of biocides.

The numerous complex trophic interactions that are likely to occur in the field, make the interpretation of population trends in these situations much more difficult than in simplified laboratory systems. Anderson (1987) recognized that the quantitative links between species and ecosystem pro-cesses could not be resolved without a more synergistic approach, i.e. analysis of whole food webs. Food web analysis requires both the measure-ment of all microbial and faunal populations and computer simulation to model flows of nutrients between the populations. There is a need for such simulations, because of the number of potential interactions, even in a very much simplified food web (Fig. 4.4). Elliott *et al.* (1988) classified the

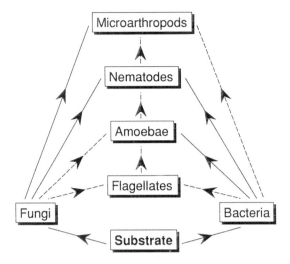

**Fig. 4.4.** Simplified representation of a below-ground food web, showing the major (solid lines) and minor (dashed lines) flows of nutrients.

decomposer organisms of a shortgrass prairie into functional groups (such as bacterial, fungal and amoebal-feeders). They also concluded that it was necessary to study the entire food web to understand the dynamics of the ecosystem, because of the complex nature of the interactions. It was estimated that bacteria would actually be responsible for the mineralization of most N (4.5 g m$^{-2}$ year$^{-1}$), but 37% of the total N mineralized could be attributed to soil fauna. Protozoa consumed 1.6 g N m$^{-2}$ year$^{-1}$ and would eliminate 1.2 g inorganic N and 90 mg organic N (as faeces) (Hunt *et al.*, 1987; Elliott *et al.*, 1988). The vast majority (98%) of N mineralized by protozoa was as a result of amoebae rather than flagellates. Fauna were more active mineralizers than bacteria, mineralizing 1400% of their biomass N per year compared with 60% for bacteria. This was because of their high turnover rates and because protozoa excrete a high fraction of the N they consume. Direct mineralization inadequately reflected the importance of the fauna, as grazers provide labile substrates which can be rapidly converted to NH$_4$ by bacteria. Thus, the calculated 37% of N mineralized by the total fauna is likely to be an underestimate.

Nitrogen budgets have similarly been calculated for four arable cropping systems in Sweden (Rosswall and Paustian, 1984). Fauna became increasingly important as the systems became poorer in N, due to enhanced immobilization by microbes as the N content of substrates decreased. Only protozoa and bacterial-feeding nematodes were estimated to excrete significant amounts of N and, following pulses of low N substrates, fauna may well be the sole mineralizers of N (Hunt *et al.*, 1987). This was reflected in the percentage of net mineralization due to the fauna under different crops; i.e. meadow fescue (19%), lucerne (20%), fertilized barley (31%) and unfertilized barley (36%). Food web studies on a Dutch polder gave similar values for the contribution of protozoa to N mineralization. Bacteria were estimated to account for approximately 67% and amoebae 18% of the total N mineralized (De Ruiter *et al.*, 1993). As in the other studies in Sweden and the USA, the amoebae were the most important faunal group. The effect of removing amoebae from the food web was calculated to greatly exceed their direct effect. Thus, while De Ruiter *et al.* (1993) calculated that 18% of N was mineralized through direct amoebal excretion, the total contribution of amoebae to N mineralization was estimated at 35% and emphasized the importance of indirect effects. A different type of analysis of the food web from the same site showed that there was a clustering of functional groups based on food source, rather than taxonomy (Moore *et al.*, 1990). Thus, the numbers of all bacterial-feeding fauna fluctuated in the same way and this particular trophic group was again dominated by amoebae. Both these Dutch studies recognized that the food web models used were sensitive to the natural death rate of protozoa and to the accuracy of the biomass estimates, which often have large coefficients of variation for protozoa.

Most of the published food web calculations are based on steady-state parameters. Verhoef and Brussaard (1990) identified the need to introduce dynamic factors (i.e. various interactions, changes in population density, seasonally changing abiotic factors, resource quality and management) to describe N supply more adequately. The inclusion of dynamic events for inferring nutrient flows was undertaken by Hunt *et al.* (1989), albeit only for soil cores incubated in the laboratory. A major difficulty in such comprehensive food web studies is to sample the soil frequently enough to trace all the major population fluctuations of all the major trophic groups. One realistic compromise is to relate detailed sampling to specific important events, such as tillage, rainfall or harvest.

## Nutrient Flows in Specific Locations in the Soil

Protozoa are primarily limited by the availability of food in the soil, so there can be little protozoan activity and protozoan-induced nutrient flow in soil without microbial growth. Protozoan effects are thus limited to those locations where there is microbial activity, i.e. the rhizosphere and sites of organic matter. The rhizosphere is, in reality, at least three different environments. Around growing root-tips, exudation is the primary source of substrate. As the root ages, cortical senescence provides the source of more nutrients and finally death of all root cells provides further substrates.

Recent experimental evidence indicates that microbial growth in the rhizo-sphere is N-limited, particularly after depletion of readily available mineral N (Breland and Bakken, 1991). At this stage, the microbial biomass could act as a significant sink of nitrogen in the rhizosphere or be rapidly converted to non-biomass soil organic N. It has been postulated that this immobilized N in the rhizosphere would be remineralized by protozoan activity (Coleman *et al.*, 1983). Some theoretical calculations have also indicated that N contained in root exudates could be entirely recycled back to the plant by grazing fauna (Griffiths and Robinson, 1992). There is, therefore, some uncertainty regarding the fate of immobilized N in the rhizosphere. Clarholm's (1985) proposal that root exudation would stimulate bacterial mineralization of soil organic N and supply some readily available N through protozoan activity is unlikely to supply significant amounts of N for crop growth. This is because of the total amount of N and C:N ratio of exudate (Griffiths and Robinson, 1992), and the limited rooting density of most arable crops (Anderson, 1987).

Dead roots, although derived from the rhizosphere, can be considered as just one of several sources of organic additions to soil. Protozoan popu-lations are known to be larger on dead roots than living roots (Christensen *et al.*, 1992). Similarly, large and rapid increases in protozoa have been observed on decomposing leaf material (Griffiths and Caul, 1993). Van Noordwijk *et al.* (1993) enumerated protozoa and bacteria on observable patches of organic matter and plant residues from a soil profile. These pat-ches contained five times more protozoan biomass than the surrounding soil, suggesting that a greater rate of nutrient cycling could occur at these locations. The larger patches of identifiable organic matter only occupied a small volume, 2–5% of the soil, so it was concluded that their contribution to overall nutrient flows would be relatively small, if not negligible. How-ever, it can be calculated that, given the steady-state conditions assumed by the authors, the patches of organic matter could supply 20% of total N mineralization (Table 4.5). Under non-steady-state conditions the popu-lation of protozoa on decomposing residues can be 50–80 times that of the surrounding soil for a limited period (Christensen *et al.* 1992; Griffiths and Caul, 1993). During these periods similar patches could supply up to 80% of soil N (Table 4.5). Although such patches may only occupy 5% or less of the soil volume, they could be significant sources of inorganic N for certain periods of crop growth.

## Conclusions

It has been demonstrated conclusively that protozoa enhance flows of nutrients in soil to the benefit of plants and microorganisms. The actual

**Table 4.5.** Estimated contribution of organic matter patches to soil N mineralization. Assuming (i) that patches occupy 5% of the soil volume; (ii) that protozoan activity in the remaining soil (95% of volume) is sufficient to supply one arbitrary unit of N from each percentage of the soil volume (i.e. 1×95% = 95 units of N); and (iii) that protozoan biomass corresponds directly with mineralizing activity.

| Increase in protozoan biomass in patch | Units of N from patch | Units of N from soil | Total N in soil | % N from patch |
|---|---|---|---|---|
| 5× (A) | 25 | 95 | 120 | 21 |
| 50× (B) | 250 | 95 | 345 | 73 |
| 80× (C) | 400 | 95 | 495 | 81 |

References: (A) Van Noordwijk *et al.*, 1993; (B) Griffiths and Caul, 1993; (C) Christensen *et al.*, 1992.

mechanism for the enhancement seems likely to be a combination of both direct and indirect effects; with current techniques it is not possible to differentiate clearly between the two effects. Estimates from the several independent food web studies confirm that protozoa, especially amoebae, are responsible for 20–40% of net N mineralization in the field. The distinction between net and gross mineralization may become important in the future. Estimates of gross rates are higher than those for net rates and inorganic N pools in soil may be completely recycled within a day (Davidson *et al.*, 1990). Estimates of net mineralization will underestimate the total amount of readily available N in circulation in the soil system. A consideration of the effects of protozoa on N turnover in laboratory systems using isotopes, as has been done for P (Barsdate *et al.*, 1974), is required. The different emphasis on P in aquatic systems and on N in soil has meant that although much is known about the fundamental aspects of protozoa/ P interactions in liquid media, there is a lack of information about these interactions in soil. The possibility that soil protozoa may enhance the availability of microbial phosphorus to plants does not appear to have been investigated. The normal experimental procedure used to determine protozoan effects is to compare treatments with and without protozoa. In real situations protozoa are always present, so a more realistic comparison may be to include a range of treatments with increasing sizes of protozoan population. A major challenge for future research will be to find ways of modifying protozoan populations, both as an experimental and a management tool, remembering the current lack of success of biocides in this respect. The biocides so far used have shown that certain food webs are dominated by fungi and others by bacteria. There is usually a good link between bacterial-feeders (protozoa and nematodes) and bacteria, but this is not always the case for fungal-feeders (nematodes and microarthropods) and fungi (Beare *et al.*, 1992). Although there are examples of fungal-feeding flagellates, amoebae and ciliates (Hekman *et al.*, 1992; Duczek, 1986; Petz *et al.*, 1985, respectively), mycophagous protozoa are usually excluded from food web calculations. There is scope for a more complete appreciation of the different protozoan trophic groups and the range of food available to protozoa, than at present.

## Acknowledgements

I would like to thank John Darbyshire for his encouragement and helpful comments. This work was funded by the Scottish Office Agriculture and Fisheries Department.

## References

Alexander, M. (1977) *Introduction to Soil Microbiology*, 2nd edn. Wiley, New York.
Anderson, J.M. (1987) Interactions between invertebrates and microorganisms: noise or necessity for soil processes. In: Fletcher, M., Gray, T.R.G. and Jones, J.G. (eds), *Ecology of Microbial Communities.* Cambridge University Press, Cambridge, pp. 125–145.
Bakken, L.R. (1985) Separation and purification of bacteria from soil. *Applied and Environmental Microbiology* 49, 1482–1487.
Bamforth, S.S. (1985) The role of protozoa in litters and soils. *Journal of Protozoology* 32, 404–409.
Band, R.N. (1959) Nutritional and related biological studies on the free-living soil amoeba, *Hartmanella rhysoides. Journal of General Microbiology*, 21, 80–95.
Barsdate, R.J., Prentki, R.T. and Fenchel, T. (1974) Phosphorus cycle of model ecosystems: significance for decomposer food chains and effect of bacterial grazers. *Oikos* 25, 239–251.
Bartholomew, W.V. (1965) Mineralization and immobilization of nitrogen in the decomposition of plant and animal residues. In: Bartholomew, W.V. and Clark, F.E. (eds), *Soil Nitrogen.* American Society of Agronomy, Madison, pp. 285–306.
Beare, M.H., Parmelee, R.W., Hendrix, P.F., Cheng, W., Coleman, D.C. and Crossley, D.A. (1992) Microbial and faunal interactions and effects on litter nitrogen and decomposition in agroecosystems. *Ecological Monographs* 62, 569–591.
Beaver, J.R. and Crisman, T.L. (1989) The role of ciliated protozoa in pelagic freshwater ecosystems. *Microbial Ecology* 17, 111–136.
Bianchi, M. (1989) Unusual bloom of star-like prosthecate bacteria and filaments as a consequence of grazing pressure. *Microbial Ecology* 17, 137–142.
Bloem, J., Starink, M., Bar-Gilissen, M-J.B. and Cappenberg, T.E. (1988) Protozoan grazing, bacterial activity, and mineralization in two-stage continuous cultures. *Applied and Environmental Microbiology* 54, 3113–3121.
Borkott, H. (1989) Elemental analyses (C,N,P,K) of invertebrate soil animals. *Zeitschrift für Pflanzenernahrung und Bodenkunde* 152, 77–80.

Bratbak, G. (1985) Bacterial biovolume and biomass estimations. *Applied and Environmental Microbiology* 49, 1488–1493.

Breland, T.A. and Bakken, L.R. (1991) Microbial growth and nitrogen immobilization in the root zone of barley (*Hordeum vulgare* L.), Italian ryegrass (*Lolium multiflorum* Lam.) and white clover (*Trifolium repens* L.). *Biology and Fertility of Soils* 12, 154–160.

Brussaard, L., Kools, J.P., Bouwman, L.A. and De Ruiter, P.C. (1991) Population dynamics and nitrogen mineralization rates in soil as influenced by bacterial grazing nematodes and mites. In: Veeresh, G.K., Rajagopal, D. and Viraktamath, C.A. (eds), *Advances in Management and Conservation of Soil Fauna*. Oxford and IBH Publishing Co. PVT Ltd, New Dehli, pp. 517–523.

Bryant, R.J., Woods, L.E., Coleman, D.C., Fairbanks, B.C., McClellan, J.F. and Cole, C.V. (1982) Interactions of bacterial and amoebal populations in soil microcosms with fluctuating moisture content. *Applied and Environmental Microbiology* 43, 747–752.

Caron, D.A. (1987) Grazing of attached bacteria by heterotrophic microflagellates. *Microbial Ecology* 13, 203–218.

Caron, D.A. and Goldman, J.C. (1988) Dynamics of protistan carbon and nitrogen cycling. *Journal of Protozoology* 35, 247–249.

Caron, D.A., Goldman, J.C., Anderson, O.K. and Dennett, M.R. (1985) Nutrient cycling in a microflagellate food chain. II Population dynamics and carbon cycling. *Marine Ecology – Progress Series* 24, 243–254.

Caron, D.A., Goldman, J.C. and Dennett, M.R. (1988) Experimental demonstration of the roles of bacteria and bacterivorous protozoa in plankton nutrient cycles. *Hydrobiologia* 159, 27–40.

Christensen, S., Griffiths, B.S., Ekelund, F. and Rønn, R. (1992) Huge increase in bacterivores on freshly killed barley roots. *FEMS Microbiology Ecology* 86, 303–310.

Clarholm, M. (1981) Protozoan grazing of bacteria in soil – impact and importance. *Microbial Ecology* 7, 343–350.

Clarholm, M. (1983) Dynamics of soil bacteria in relation to plants, protozoa and inorganic nitrogen. PhD Thesis, Swedish University of Agricultural Sciences, Uppsala.

Clarholm, M. (1985) Interactions of bacteria, protozoa and plants leading to mineralization of soil nitrogen. *Soil Biology and Biochemistry* 17, 181–188.

Clarholm, M. (1989) Effects of plant–bacterial–amoebal interactions on plant uptake of nitrogen under field conditions. *Biology and Fertility of Soils* 8, 373–378.

Cole, C.V., Elliott, E.T., Hunt, H.W. and Coleman, D.C. (1978) Trophic interactions in soils as they affect energy and nutrient dynamics. V. Phosphorus transformations. *Microbial Ecology* 4, 381–387.

Coleman, D.C., Cole, C.V., Anderson, R.V., Blaha, M., Campion, M.K., Clarholm, M., Elliott, E.T., Hunt, H.W., Shaefer, B. and Sinclair, J. (1977) An analysis of rhizosphere–saprophage interactions in terrestrial ecosystems. *Ecological Bulletins (Stockholm)* 25, 299–309.

Coleman, D.C., Anderson, R.V., Cole, C.V., Elliott, E.T., Woods, L. and Campion, M.K. (1978) Trophic interactions in soils as they affect energy and nutrient dynamics. IV. Flows of metabolic and biomass carbon. *Microbial Ecology* 4, 373–380.

Coleman, D.C., Reid, C.P.P. and Cole, C.V. (1983) Biological strategies of nutrient cycling in soil systems. *Advances in Ecological Research* 13, 1–55.

Curds, C.R. and Vandyke, J.M. (1966) The feeding habits and growth rates of some fresh-water ciliates found in activated-sludge plants. *Journal of Applied Ecology* 3, 127–137.

Cutler, D.W. and Bal, D.V. (1926) Influence of protozoa on the process of nitrogen fixation by *Azotobacter chroococcum*. *Annals of Applied Biology* 13, 516–534.

Darbyshire, J.F. (1972) Nitrogen fixation by *Azotobacter chroococcum* in the presence of *Colpoda steini* – 1. The influence of temperature. *Soil Biology and Biochemistry* 4, 359–369.

Darbyshire, J.F. and Greaves, M.P. (1967) Protozoa and bacteria in the rhizosphere of *Sinapis alba* L., *Trifolium repens* L. and *Lolium perenne* L. *Canadian Journal of Microbiology* 13, 1057–1068.

Darbyshire, J.F., Griffiths, B.S., Davidson, M.S. and McHardy, W.J. (1989) Ciliate distribution amongst soil aggregates. *Revue d'Ecologie et de Biologie du Sol* 26, 47–56.

Davidson, E.A., Stark, J.M. and Firestone, M.K. (1990) Microbial production and consumption of nitrate in an annual grassland. *Ecology* 71, 1968–1975.

De Ruiter, P.C., Moore, J.C., Zwart, K.B., Bouwman, L.A., Hassink, J., Bloem, J., De Vos, J.A., Marinissen, J.C.Y., Didden, W.A.M., Lebbink, G. and Brussaard, L. (1993) Simulation of nitrogen mineralization in the below-ground food webs of two winter wheat fields. *Journal of Applied Ecology* 30, 95–106.

Doyle, W.L. and Harding, J.P. (1937) Quantitative studies on the ciliate *Glaucoma*. Excretion of ammonia. *Journal of Experimental Biology*, 14, 462–469.

Duczek, L.J. (1986) Populations in Saskatchewan soils of spore perforating amoebae and an amoeba (*Thecamoeba granifera* s.sp *minor*) which feeds on hyphae of *Cochliobolus sativus*. *Plant and Soil* 92, 295–298.

Elliott, E.T., Coleman, D.C. and Cole, C.V. (1979) The influence of amoebae on the uptake of nitrogen by plants in a gnotobiotic soil. In: Harley, J.L. and Scott Russell, R. (eds), *The Soil-Root Interface*. Academic Press, London, pp. 221–229.

Elliott, E.T., Anderson, R.V., Coleman, D.C. and Cole, C.V. (1980) Habitable pore space and microbial trophic interactions. *Oikos* 35, 327–335.

Elliott, E.T., Horton, K., Moore, J.C., Coleman, D.C. and Cole, C.V. (1984) Mineralization dynamics in fallow dryland wheat plots, Colorado. *Plant and Soil* 76, 149–155.

Elliott, E.T., Hunt, H.W. and Walter, D.E. (1988) Detrital food-web interactions in North American grassland ecosystems. *Agriculture, Ecosystems and Environment* 24, 41–56.

Fenchel, T. (1982) Ecology of heterotrophic microflagellates. IV. Quantitative occurrence and importance as bacterial consumers. *Marine Ecology – Progress Series* 9, 35–42.

Fenchel, T. and Harrison, P. (1976) The significance of bacterial grazing and mineral cycling for the decomposition of particulate detritus. In: Anderson, J.M. and MacFadyen, A. (eds), *The Role of Terrestrial and Aquatic Organisms in Decomposition Processes*. Blackwell Scientific Publications, Oxford, pp. 285–299.

Frey, J.S., McClellan, J.F., Ingham, E.R. and Coleman, D.C. (1985) Filter-out-grazers (FOG): A filtration experiment for separating protozoan grazers in soil. *Biology and Fertility of Soils* 1, 73–79.

Gel'tser, J.G. (1991) Free-living protozoa as a component of soil biota. *Soil Biology* (translated from *Pochvovedeniye*) 8, 66–79.

Griffiths, B.S. (1986) Mineralization of nitrogen and phosphorus by mixed cultures of the ciliate protozoan *Colpoda steinii*, the nematode *Rhabditis* sp. and the bacterium *Pseudomonas fluorescens*. *Soil Biology and Biochemistry* 18, 637–641.

Griffiths, B.S. (1989a) Enhanced nitrification in the presence of bacteriophagous protozoa. *Soil Biology and Biochemistry* 21, 1045–1051.

Griffiths, B.S. (1989b) The role of bacterial feeding nematodes and protozoa in rhizosphere nutrient cycling. *Aspects of Applied Biology* 22, 141–145.

Griffiths, B.S. (1990) A comparison of microbial feeding nematodes and protozoa in the rhizosphere of different plants. *Biology and Fertility of Soils* 9, 83–88.

Griffiths, B.S. and Caul, S. (1993) Migration of bacterial-feeding nematodes, but not protozoa, to decomposing grass residues. *Biology and Fertility of Soils* 15, 201–207.

Griffiths, B.S. and Robinson, D. (1992) Root-induced nitrogen mineralization: A nitrogen balance model. *Plant and Soil* 139, 253–263.

Griffiths, B.S., Ekelund, F., Rønn, R. and Christensen, S. (1993) Protozoa and nematodes on decomposing barley roots. *Soil Biology and Biochemistry* 25, 1293–1295.

Güde, H, (1979) Grazing by protozoa as selection factor for activated sludge protozoa. *Mircrobial Ecology* 5, 225–237.

Güde, H. (1985) Influence of phagotrophic processes on the regeneration of nutrients in two-stage continuous culture systems. *Microbial Ecology* 11, 193–204.

Hattori, T. (1989) Soil aggregates as microhabitats of microorganisms. *The Reports of the Institute for Agricultural Research Tohoku University* 37, 23–36.

Heal, O.W. and Felton, M.J. (1970) Soil amoebae: their food and their reaction to microflora exudates. In: Watson, A. (ed.), *Animal Populations in Relation to their Food Resources.* Blackwell Scientific Publications, Oxford, pp. 145–162.

Heijnen, C.E. and Van Veen, J.A. (1991) A determination of protective microhabitats for bacteria introduced into soil. *FEMS Microbiology Ecology* 85, 73–80.

Hekman, W.E., Van den Boogert, P.J.H.F. and Zwart, K.B. (1992) The physiology and ecology of a novel, obligate mycophagous flagellate. *FEMS Microbiology Ecology* 86, 255–265.

Hunt, H.W., Cole, C.V., Klein, D.A. and Coleman, D.C. (1977) A simulation model for the effect of predation on bacteria in continuous culture. *Microbial Ecology* 3, 257–278.

Hunt, H.W., Coleman, D.C., Ingham, E.R., Ingham, R.E., Elliott, E.T., Moore, J.C., Rose, S.L., Reid, C.P.P. and Morley, C.R. (1987) The detrital food web in a shortgrass prairie. *Biology and Fertility of Soils* 3, 57–68.

Hunt, H.W., Elliott, E.T. and Walter, D.E. (1989) Inferring trophic transfers from pulse-dynamics in detrital food webs. *Plant and Soil* 115, 247–259.

Ingham, E.R. (1985) Review of the effects of 12 selected biocides on target and non-target soil organisms. *Crop Protection* 4, 3–32.

Ingham, E.R. and Coleman, D.C. (1984) Effects of streptomycin, cycloheximide, fungizone, captan, carbofuran, cygon and PCNB on soil microorganisms. *Microbial Ecology* 10, 345–358.

Ingham, E.R. and Horton, K.A. (1987) Bacterial, fungal and protozoan responses to chloroform fumigation in stored soil. *Soil Biology and Biochemistry* 19, 545–550.

Ingham, E.R., Trofymow, J.A., Ames, R.N., Hunt, H.W., Morley, C.R., Moore, J.C. and Coleman, D.C. (1986a) Trophic interactions and nitrogen cycling in a semi-arid grassland soil. I. Seasonal dynamics of the natural populations, their interactions and effects on nitrogen cycling. *Journal of Applied Ecology* 23, 597–614.

Ingham, E.R., Cambardella, C. and Coleman, D.C. (1986b) Manipulation of bacteria, fungi and protozoa by biocides in lodgepole pine forest microcosms: Effects on organism interactions and nitrogen mineralization. *Canadian Journal of Soil Science* 66, 261–272.

Ingham, E.R., Trofymow, J.A., Ames, R.N., Hunt, H.W., Morley, C.R., Moore, J.C. and Coleman, D.C. (1986c) Trophic interactions and nitrogen cycling in a semi-arid grassland soil. II. System responses to removal of different groups of soil microbes or fauna. *Journal of Applied Ecology* 23, 615–630.

Ingham, E.R., Coleman, D.C. and Moore, J.C. (1989) An analysis of food-web structure and function in a shortgrass prairie, a mountain meadow, and a lodgepole pine forest. *Biology and Fertility of Soils* 8, 29–37.

Ingham, E.R., Parmelee, R., Coleman, D.C. and Crossley, D.A. (1991) Reduction of microbial and faunal groups following application of streptomycin and captan in Georgia no-till agroecosystems. *Pedobiologia* 35, 297–304.

Johannes, R.E. (1965) Influence of marine protozoa on nutrient regeneration. *Limnology and Oceanography* 10, 434–442.

Johannes, R.E. (1968) Nutrient regeneration in lakes and oceans. In: Droop, M.R. and Ferguson Wood, E.J. (eds), *Advances in Microbiology of the Sea*, Vol. 1. Academic Press, London, pp. 203–213.

Killham, K., Amato, M. and Ladd, J.N. (1993) Effect of substrate localisation in soil and soil pore-water regime on carbon turnover. *Soil Biology and Biochemistry* 25, 57–62.

Kuikman, P.J. (1990) Mineralization of nitrogen by protozoan activity in soil. PhD Thesis, Agricultural University of Wageningen, The Netherlands.

Kuikman, P.J. and Van Veen, J.A. (1989) The impact of protozoa on the availability of bacterial nitrogen to plants. *Biology and Fertility of Soils* 8, 13–18.

Kuikman, P.J., Van Vuuren, M.M.I. and Van Veen, J.A. (1989) Effect of soil moisture regime on predation by protozoa of bacterial biomass and the release of bacterial nitrogen. *Agriculture, Ecosystems and Environment* 27, 271–279.

Kuikman, P.J., Jansen, A.G., Van Veen J.A. and Zehnder, A.J.B. (1990a) Protozoan predation and the turnover of soil organic carbon and nitrogen in the presence of plants. *Biology and Fertility of Soils* 10, 22–28.

Kuikman, P.J., Van Elsas, J.D., Jansen, A.G., Burgers, S.L.G.E. and Van Veen, J.A. (1990b) Population dynamics and activity of bacteria and protozoa in relation to their spatial distribution in soil. *Soil Biology and Biochemistry* 22, 1063–1073.

Kuikman, P.J., Jansen, A.G. and Van Veen, J.A. (1991) $^{15}$N-nitrogen mineralization from bacteria by protozoan grazing at different soil moisture regimes. *Soil Biology and Biochemistry* 23, 193–200.

Lawrie, N.R. (1935) Studies in the metabolism of protozoa. I. The nitrogenous metabolism and respiration of *Bodo caudatus*. *Biochemical Journal* 29, 588–598.

Lee, S. and Fuhrman, J.A. (1987) Relationships between biovolume and biomass of naturally derived marine bacterioplankton. *Applied and Environmental Microbiology* 53, 1298–1303.

Levrat, P. (1990) Contribution a l'étude des interactions entre protozoaires et microflore du sol: effet d'une amibe bactériophage *Acanthamoeba castellanii* sur le métabolism de *Pseudomonas* fluorescents. PhD Thesis, University Claude Bernard – Lyon I, France.

Levrat, P., Pussard, M. and Alabouvette, C. (1989) Action d'*Acanthamoeba castellanii* (Protozoa, Amoebida) sur la production de sidérophores par la bactérie *Pseudomonas putida*. *Comptes Rendus hebdomadaire des seances de l'Académie des Sciences, Paris* 308 series III, 161–164.

Levrat, P., Pussard, M. and Alabouvette, C. (1992) Enhanced bacterial metabolism of a *Pseudomonas* strain in response to the addition of culture filtrate of a bacteriophagous amoeba. *European Journal of Protistology* 28, 79–84.

Lousier, J.D. and Bamforth, S.S. (1988) Soil protozoa. In: Dindal, D.L. (ed.), *Soil Biology Guide.* John Wiley and Sons, London, pp. 97–136.

Meiklejohn, J. (1930) The relation between the numbers of a soil bacterium and the ammonia produced by it in peptone solutions; with some reference to the effect on this process of the presence of amoebae. *Annals of Applied Biology* 17, 614–637.

Meiklejohn, J. (1932) The effect of *Colpidium* on ammonia production by soil bacteria. *Annals of Applied Biology* 19, 584–608.

Moore, J.C., Zwetslooth, H.J.C. and De Ruiter, P.C. (1990) Statistical analysis and simulation modelling of the below ground food webs of two winter wheat management practices. *Netherlands Journal of Agricultural Science* 38, 303–316.

Nasir, S.M. (1923) Investigations on the relationship of protozoa to soil fertility with special reference to nitrogen fixation. *Annals of Applied Biology* 10, 122–133.

Nilsson, J.R. (1987) Structural aspects of digestion of *Escherichia coli* in *Tetrahymena. Journal of Protozoology* 34, 1–6.

Odum, E.P. (1971) *Fundamentals of Ecology*, 3rd edn. W.B. Saunders Company, Philadelphia.

Petz, W., Foissner, W. and Adam, H. (1985) Culture, food selection and growth rate in the mycophagous ciliate *Grossglockneria acuta* Foissner, 1980: First evidence of autochthonous soil ciliates. *Soil Biology and Biochemistry* 17, 871–875.

Porter, K.G., Sherr, E.B., Sherr, B.F., Pace, M. and Sanders, R.W. (1985) Protozoa in planktonic food webs. *Journal of Protozoology* 32, 409–415.

Postma, J. and Van Veen, J.A. (1990) Habitable pore space and survival of *Rhizobium leguminosarum* biovar. *trifolii* introduced into soil. *Microbial Ecology* 19, 149–161.

Pussard, M. and Rouelle, J. (1986) Predation of the microflora. Action of protozoa on the dynamics of bacterial populations. *Protistologica* 22, 105–110.

Riemann, B., Jorgensen, N.O.G., Lampert, W. and Fuhrman, J.A. (1986) Zooplankton induced changes in dissolved free amino-acids and production rates of freshwater bacteria. *Microbial Ecology* 12, 247–258.

Ritz, K. and Griffiths, B.S. (1987) Effects of carbon and nitrate additions to soil upon leaching of nitrate, microbial predators and nitrogen uptake by plants. *Plant and Soil* 102, 229–237.

Rosswall, T. and Paustian, K. (1984) Cycling of nitrogen in modern agricultural systems. *Plant and Soil* 76, 3–21.

Rutherford, P.M. and Juma, N.G. (1992a) Influence of soil texture on protozoan-induced mineralization of bacterial carbon and nitrogen. *Canadian Journal of Soil Science* 72, 183–200.

Rutherford, P.M. and Juma, N.G. (1992b) Simulation of protozoan-induced mineralization of bacterial carbon and nitrogen. *Canadian Journal of Soil Science* 72, 201–216.

Schönborn, W. (1992) The role of protozoan communities in freshwater and soil ecosystems. *Acta Protozoologica* 31, 11–18.

Sherr, B.F., Sherr, E.B. and McDaniel, J. (1992) Effect of protistan grazing on the frequency of dividing cells in bacterioplankton assemblages. *Applied and Environmental Microbiology* 58, 2381–2385.

Shikano, S., Luckinbill, L.S. and Kurihara, Y. (1990) Changes of traits in a bacterial population associated with protozoan predation.*Microbial Ecology* 20, 75–84.

Sibbald, M.J. and Albright, L.J. (1988) Aggregated and free bacteria as food sources for heterotrophic microflagellates. *Applied and Environmental Microbiology* 54, 613–616.

Sieburth, J.M. and Davis, P.G. (1982) The role of heterotrophic nanoplankton in the grazing and nuturing of planktonic bacteria in the Sargasso and Caribbean Sea. *Annales d l'Institut Oceanographique* 58 (Supplement), 285–296.

Sinclair, J.L., McClellan, J.F. and Coleman, D.C. (1981) Nitrogen mineralization by *Acanthamoeba polyphaga* in grazed *Pseudomonas paucimobilis* populations. *Applied and Environmental Microbiology* 42, 667–671.

Singh, B.N. (1941) Selectivity in bacterial food by soil amoebae in pure and mixed culture and in sterilised soil. *Annals of Applied Biology* 28, 52–65.

Singh B.N. (1942) Selection of bacterial food by soil flagellates and amoebae. *Annals of Applied Biology* 29, 18–22.

Singh, B.N. (1964) Soil protozoa and their probable role in soil fertility. *Bulletin of the National Institute of Sciences of India* 26, 238–244.

Sleigh, M.A. (1989) *Protozoa and Other Protists.* Edward Arnold, London.

Sohlenius, B. (1968) Influence of microorganisms and temperature upon some Rhabditid nematodes. *Pedobiologia* 8, 137–145.

Stout, J.D. (1980) The role of protozoa in nutrient cycling and energy flow. *Advances in Microbial Ecology* 4, 1–50.

Stout, J.D. and Heal, O.W. (1967) Protozoa. In: Burgess, A. and Raw, F. (eds), *Soil Biology.* Academic Press, London, pp. 149–196.

Taylor, W.D. (1986) The effect of grazing by a ciliated protozoan on phosphorus limitation of heterotrophic bacteria in batch culture. *Journal of Protozoology* 33, 47–52.

Taylor, W.D. and Berger, J. (1976) Growth responses of cohabiting ciliate protozoa to various prey bacteria. *Canadian Journal of Zoology* 54, 1111–1114.

Telegedy-Kovats, L. de (1932) The growth and respiration of bacteria in sand cultures in the presence and absence of protozoa. *Annals of Applied Biology* 19, 65–86.

Van Noordwijk, M., De Ruiter, P.C., Zwart, K.B., Bloem, J., Moore, J.C., Van Faassen, H.G. and Burgers, S.L.G.E. (1993) Synlocation of biological activity, roots, cracks and recent organic inputs in a sugar-beet field. *Geoderma* 56, 265–276.

Van Veen, J.A. and Paul, E.A. (1979) Conversion of biovolume measurements of soil organisms grown under various moisture tensions, to biomass and their nutrient content. *Applied and Environmental Microbiology* 37, 686–692.

Vargas, R. (1990) Soil microbiology advances: protozoa and their importance in nitrogen mineralization. *Agronomia Costarricense* 14, 121–143.

Verhagen, F.J.M. (1992) Nitrification versus immobilization of ammonium in grassland soils and effects of protozoan grazing. PhD Thesis, University of Nijmegen, The Netherlands.

Verhagen, F.J.M. and Laanbroek, H.J. (1992) Effects of grazing by flagellates on competition for ammonium between nitrifying and heterotrophic bacteria in chemostats. *Applied and Environmental Microbiology* 58, 1962–1969.

Verhoef, H.A. and Brussaard, L. (1990) Decomposition and nitrogen mineralization in natural and agroecosystems: The contribution of soil animals. *Biogeochemistry* 11, 175–211.

Visser, S. (1985) Role of the soil invertebrates in determining the composition of soil microbial communities. In: Fitter, A.H., Atkinson, D., Read, D.J. and Usher, M.B. (eds), *Ecological Interactions in Soil.* Blackwell Scientific Publications, Oxford, pp. 297–317.

Woods, L.E., Cole, C.V., Elliott, E.T., Anderson, R.V. and Coleman, D.C. (1982) Nitrogen transformations in soil as affected by bacterial-microfaunal interactions. *Soil Biology and Biochemistry* 14, 93–99.

Zwart, K.B. and Brussaard, L. (1991) Soil fauna and cereal crops. In: Firbank, L.G., Carter, N., Darbyshire, J.F. and Potts, G.R. (eds), *The Ecology of Temperate Cereal Fields.* Blackwell Scientific Publications, Oxford, pp. 139–168.

Zwart, K.B. and Darbyshire, J.F. (1992) Growth and nitrogenous excretion of a common soil flagellate *Spumella* sp. – a laboratory experiment. *Journal of Soil Science* 43, 145–157.

# Rhizosphere Protozoa: Their Significance in Nutrient Dynamics

K.B. Zwart[1], P.J. Kuikman[2] and J.A. Van Veen[3]
[1]DLO-Research Institute for Agrobiology and Soil Fertility,
Oosterweg 92, PO Box 129, 9750 AC Haren, The Netherlands:
[2]DLO-Research Institute for Agrobiology and Soil Fertility,
Bornesteeg 65, PO Box 14, 6700 AA Wageningen, The
Netherlands: [3]DLO-Research Institute for Plant Protection,
Binnenhaven 12, PO Box 9060, 6700 GW Wageningen, The
Netherlands.

## Introduction

Compared with aquatic protozoa, soil protozoa, particularly those residing near plant roots, have been relatively overlooked by scientists. This neglect is surprising, because in most soils intense microbial activity can be expected in the small zone of soil adjacent to plant roots, i.e. the *rhizosphere*. As protozoa feed mainly on bacteria, yeasts or fungi, the highest protozoan activity might be expected in this zone too, but there are only a small number of papers dealing with rhizosphere protozoa. Foissner (1987), in his extensive review of soil protozoan literature, did not even devote a section to the subject. In this chapter, we discuss the qualitative and quantitative aspects of rhizosphere protozoa, especially in relation to their metabolism and the release of nutrients to plant roots. Because the activity of protozoa is largely dependent on microfloral activity, we also discuss the growth of other microorganisms in the rhizosphere. The relationships of microbial growth to the carbon and energy contents in the root exudates are emphasized. In addition, the effects of rhizosphere protozoa on plant roots and diseases are discussed briefly.

# Dimensions of the Rhizosphere

The rhizosphere, as originally defined by Hiltner (1904), is the narrow zone of soil surrounding plant roots and it contains very large microbial populations stimulated by root activity. The inner and outer boundaries of the rhizosphere are poorly demarcated (Darbyshire and Greaves, 1973). The rhizosphere dimensions vary with plant species and cultivar, the stage of plant development and the type of soil in which the plant is growing. Rovira (1969) regarded a zone of 1–2 mm around roots as the zone in which microbial growth was influenced by root exudates. In sterile soil with maize, root exudates may diffuse over a distance of 20 mm (Helal and Sauerbeck, 1986). The length of root hairs may also extend the rhizosphere significantly. In two computer simulations of the rhizosphere (Newman and Watson, 1977; Darrah, 1991a,b), the outer boundary influenced by soluble root exudates was set at 2000 μm and this is the value used in calculations for the rhizosphere radius in this chapter.

The metabolic activity and growth of the root itself can create gradients with respect to water, oxygen, carbon dioxide, pH, redox potential, organic matter and mineral nutrients in the rhizosphere. Gradients in soil water content close to the root are generally small, unless the soil is dry or shows a rapid decrease in hydraulic conductivity when drying, e.g. coarse sand. Gradients in nutrient concentration depend on uptake activity, soil water content and the apparent adsorption constant. For nitrate the gradients are small, for ammonium and potassium intermediate and for phosphate large (Nye and Tinker, 1977; De Willigen and Van Noordwijk, 1987). These gradients affect microbial activity, and in turn the microbes can also modify rhizosphere conditions (Römheld, 1990). Clearly, the rhizosphere is a very dynamic environment with individual roots growing into a given soil region, differentiating, decomposing and finally disappearing from the area, all within a short period of time (Clarholm, 1985).

# Root Exudates

The intensified microbial activity that is observed in the rhizosphere is the result of the release of dissolved and particulate organic components from plant roots. Whipps (1990) identified four classes of chemical compounds in root exudates: water-soluble exudates (e.g. sugars, amino acids, organic acids, hormones and vitamins) that diffuse out of roots, polymeric carbohydrates (that are actively secreted by plants at the expense of metabolic energy), lysates (released by autolysis) and gases (such as ethylene and $CO_2$). Most of these compounds can be utilized by a large variety of soil microorganisms as a source of carbon, nitrogen and energy (Newman, 1978;

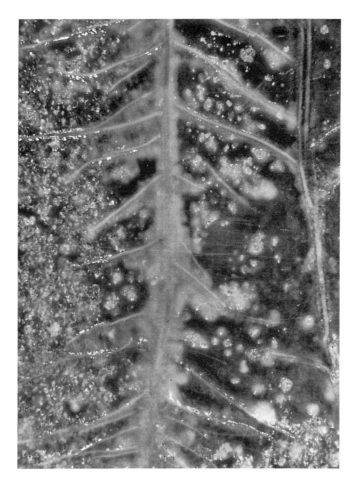

**Fig. 5.1.** Growth of bacteria on agar as a result of root exudation. After contact with the soil/root surface of a maize plant (*Zea mays* L.), the agar was subsequently incubated at 25°C (Römheld, 1990).

Elliott *et al.*, 1984). Römheld (1990) aptly demonstrated the stimulatory effect of maize exudates on microbial growth (Fig. 5.1). The amounts of carbon and nitrogen released by roots vary between plant species and according to conditions (Rovira, 1969).

## Carbonaceous exudates

Beck and Gilmour (1983) found that 3–3.7% of $^{14}C$-labelled photosynthate of wheat plants grown in liquid cultures was released as soluble exudates. Merckx *et al.* (1987) found that 2–4% of the total $^{14}C$ fixed by maize plants was retained as a microbial residue in soil after 42 days of growth. Martens (1990) observed that 21.1–41% of the total $^{14}C$ fixed by maize was translocated below ground and that approximately half of this was released as $CO_2$ by the roots and microorganisms, while only 4–5% was found as soil residual carbon. Cheshire and Mundie (1990) reported that 25% of the $^{14}C$-labelled carbon exuded by maize plants in soil was soluble and this was approximately 5% of the label in the total plant–soil system. Whipps (1990) compared the relative amounts of photosynthate C transferred to roots and released into the soil when several plant species were grown in pots (Table 5.1). Such translocations from the shoots to the roots varied from 41 to 80% of the net fixed carbon. In general, more than 50% of the carbon translocated to the roots was lost as exudates from living roots, of which 12–59% was released as $CO_2$ during respiration from roots or microorganisms. The remaining 13–43% was accumulated in the soil as dissolved or

**Table 5.1.** Percentage of net fixed $^{14}C$ translocated to the roots and the percentage of $^{14}C$ in root exudates when plant shoots were exposed continuously to $^{14}C$ and plants were grown in soil; after Whipps (1990).

| Plant | % of net photosynthate $^{14}C$ transferred to the roots[a] | % of root $^{14}C$ lost as exudates | | |
|---|---|---|---|---|
| | | Total[b] | Rhizodeposition in soil[c] | Respiration |
| Wheat | 45 | 50 | 13 | 37 |
| Barley | 41 | 56 | 27 | 29 |
| Maize | 35 | 58 | 28 | 30 |
| Mustard | na | 73 | 14 | 59 |
| Tomato | 42 | 81 | 43 | 38 |
| Pea | 55 | 78 | 36 | 42 |
| Blue grama grass | 80 | 51 | 39 | 12 |

[a] ($\mu$Ci in root + rhizodeposition + $CO_2$ from root and soil microorganisms) / total $\mu$Ci fixed by plant × 100.
[b] ($\mu$Ci in $CO_2$ from root and soil microorganisms + rhizodeposition in soil) / ($\mu$Ci in root + rhizodeposition in soil + $CO_2$ from root and soil microorganisms) × 100.
[c] ($\mu$Ci in rhizodeposition in soil) / ($\mu$Ci in root + rhizodeposition in soil + $CO_2$ from root and soil microorganisms) × 100.
($\mu$Ci = amount of $^{14}C$).
na: data not available.

particulate plant material, microorganisms or other organisms. Christiansen-Weniger *et al.* (1992) observed that aluminium-tolerant wheat cultivars exuded significantly more carbon than aluminium-sensitive plants (53 and 36%, respectively, of the total root production).

Swinnen and Van Veen (unpublished results) used pulse-labelling of plants with $^{14}CO_2$ to estimate the total carbon input into the soil for spring wheat during the growing season under field conditions. During an entire growing season, 1900 kg C $ha^{-1}$ was translocated below ground and at harvest 400 kg was present in the roots. Of the remaining 1500 kg, 66% was respired by roots and microorganisms and 33% was released as dissolved or particulate material. Whipps (1990) collated root exudation data from field experiments of several authors. For wheat, the total annual rhizodeposition, i.e. root C + soil C + $CO_2$ C, ranged between 1.2 and 2.9 tonnes of C $ha^{-1.}$ During an entire growing season, 15% of the net fixed C was transferred to the roots and 50% of this C was lost as $CO_2$ and 25% was deposited in the soil. For prairie grasses, the total annual rhizodeposition was 1.3–2.1 tonnes of C $ha^{-1}$, while 35–80% of the total net fixed C was translocated to the roots, depending on the plant species present. Fogel and Hunt (1983) estimated that between 40 and 73% of the net fixed C was transferred by trees to the roots and that 30–90% of this C was lost to the plant, i.e. between 5.8 and 7.5 tonnes of C $ha^{-1}$ $year^{-1.}$ Carbon input through rhizodeposition is generally believed to be more significant in perennial than in annual plants. The production of non-$CO_2$ C by roots of different plants $ha^{-1}$ soil is presented in Table 5.2. In summary, 1–5 tonnes C $ha^{-1}$ $year^{-1}$ are lost from the roots. Assuming steady-state conditions in the pool of soil organic matter an equivalent quantity of C will be metabolized by microorganisms. The highest rates of rhizodeposition of C are believed to occur near the growing root-tip (Trofymow *et al.*, 1987; Darrah, 1991b). Darrah (1991b) published a simulation study of the dynamics of exuded carbon and rhizosphere microbial populations near the growing root-tip for two possible lengths of the exudation zone immediately behind the root-tip for a root elongating at 0.04 cm $h^{-1}$. If this exudation zone was 0.5 cm long, a peak of carbonaceous exudates, extending across the entire rhizosphere, would move down the soil cylinder associated with the growing root-tip. Behind this peak and approximately 150 h after the root-tip had reached a given soil layer, the concentrations of soluble C rapidly decreased, because of the rapidly proliferating microbial biomass that was growing at a maximum specific growth rate of 0.165 cm $h^{-1}$. If the exuding zone was 5 cm, the wave of carbon was much more extensive, but the amount of microbial biomass produced and its production rate remained approximately the same as when the exudation zone was 0.5 cm.

The magnitude of microbial respiration in the rhizosphere, based on calculations of root exudation of $^{14}C$, is unresolved even when the net $^{14}CO_2$ fixation by plants is known. Johnen and Sauerbeck (1977) assumed that

**Table 5.2.** Total amount of non-$CO_2$ carbon released by different plant species (in mg g$^{-1}$ dry root or kg C ha$^{-1}$ year $^{-1}$). Root lengths according to Van Noordwijk and Brouwer (1991).

| Plant | Root length m ha$^{-1}$ (10$^8$) | C loss by roots | | Reference |
| | | mg g$^{-1}$ root | kg ha$^{-1}$ | |
|---|---|---|---|---|
| Wheat | 250 | | 500 | Swinnen and Van Veen (unpublished) |
| | | | 300–525 | Whipps (1990) |
| | | | 325 | Keith et al. (1986) |
| | | 34 | 42.5 | Whipps and Lynch (1983) |
| Barley | 80 | | 190–1390 | Clarholm (1985) |
| | | | 900–3000 | Wood (1987) |
| | | 2845 | 1130 | Wood (1987) |
| | | 34 | 13.6 | Barber and Lynch (1977) |
| | | 75 | 30 | Whipps and Lynch (1983) |
| | | 45–82 | 17–33 | Whipps and Lynch (1983) |
| | | 15 | 6 | Newman and Watson (1977) |
| Maize | 94 | 6–60 | 0.3–30 | Newman and Watson (1977) |
| Agropyron cristatum | 630 | 43.5 | 136 | Klein et al. (1988) |
| A. smithii | 630 | 26.6 | 84 | Klein et al. (1988) |
| Bouteloua gracilis | 630 | 46.8 | 147 | Klein et al. (1988) |
| Lucerne | 250 | 247 | 309 | Newman and Watson (1977) |
| Robinia pseudoacacia | 250 | 38 | 47.5 | Newman and Watson (1977) |
| Pinus sp. | 250 | 67–244 | 84–305 | Newman and Watson (1977) |

30% of the $CO_2$ lost from roots of wheat and mustard was derived from root respiration and the remaining 70% represented microbial respiration. Swinnen and Van Veen (unpublished results), on the other hand, calculated from their plant exudate-labelling experiments that 93% of the total $CO_2$ was produced by roots with only 7% from microbial activity.

### Nitrogenous exudates

Plants exude nitrogen mostly as proteins, peptides and amino acids (Vancura, 1988; Table 5.3). The highest rates of N exudation occur near the growing root-tip (Trofymow et al., 1987). In four plant species, barley, wheat, cucumber and garden bean, the exudates contained 2–5% nitrogen

**Table 5.3.** Nitrogen in root exudates (Vancura, 1988). For the C:N ratio, a 50% C content in the exudate was assumed.

| Plant | Total N mg g$^{-1}$ | Amino acids mg N g$^{-1}$ | Amino acids mg g$^{-1}$ | Proteins mg N g$^{-1}$ | Proteins mg g$^{-1}$ | Residual N mg g$^{-1}$ | C:N |
|---|---|---|---|---|---|---|---|
| Barley | 24.0 | 7.9 | 49.1 | 11.2 | 69.7 | 4.9 | 20.9 |
| Wheat | 22.8 | 13.0 | 81.3 | 9.8 | 61.3 | 0 | 21.9 |
| Cucumber | 44.6 | 25.7 | 160.6 | 18.9 | 118.1 | 0 | 11.2 |
| Garden bean | 33.0 | 15.9 | 99.2 | 6.6 | 41.5 | 10.5 | 15.2 |

by weight. If the C in exudates is assumed to be 42%, the C:N ratio in root exudates will be approximately 14. Boulter *et al.* (1966) found that ammonia (25%), homo-serine (23%), aspartic acid (12%) and glutamic acid (10%) were the major components of the exudates when pea seedlings were grown in quartz sand. These results, however, represent a much lower release of total N than that reported by Vancura (1988). Matsumoto *et al.* (1979) found that glumatic acid accounted for 60% of the amino acids exuded by maize roots, followed by alanine. In non-sterile model systems, Klein *et al.* (1988) reported surprisingly high N concentrations of exuded amino acids, sugars and organic acids from three different grasses resulting in C:N ratios in the exudates of between 2.0 and 2.7, i.e. even lower than such ratios for most amino acids. Biondini *et al.* (1988) reported C:N ratios of 14 for *Bouteloua gracilis* and *Agropyron smithii* (dominant grasses in a shortgrass steppe ecosystem) and 15.6 for *A. cristatum* (the dominant grass in a crested wheatgrass system). Trofymow *et al.* (1987) reported a C:N ratio of 21 in rhizodeposition of axenic oats. More recently, Janzen (1990) investigated the total deposition of N by wheat roots in soil, after a short pulse of $^{15}$N-labelled ammonia was applied to the shoots. He found that three plants exuded 26 and 76 mg N at low and high N fertility treatments, representing 18 and 33%, respectively, of the total N yield of the wheat plants. Moreover, in the low N fertility treatment, most of the exuded N was organic. This indicates that the C to N ratio of the root exudates was similar to that of the plant itself. From the data reported in the literature, one may conclude that the C to N ratio of root exudates may vary, although the amount of N lost by plants seems to be considerable in all instances. For many plants, the C:N ratio of their root exudates may be close to that of the plant itself.

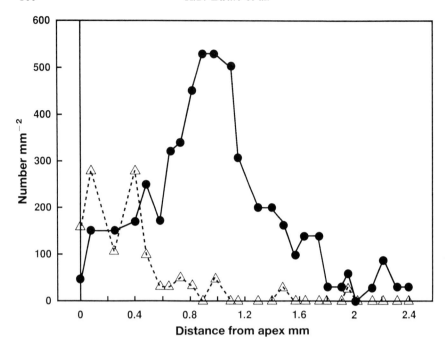

**Fig. 5.2.** The density of active (△) and of cystic (•) amoebae on the roots of wheat growing on agar as a function of the root length (from Coûteaux *et al.*, 1988).

## Rhizosphere Populations of Protozoa

Although Foster and Dormaar (1991) have published electron micrographs of rhizosphere soil with amoebae and Coûteaux *et al.* (1988) showed that large numbers of active amoebae occur near the root-tips when plants are grown on agar (Fig. 5.2), relatively few estimates of the number of protozoa present in the rhizosphere soil have been published (Table 5.4). Detailed comparisons of these sets of data are handicapped by the many different methods used to estimate protozoan numbers and the many different environmental conditions involved. For example, many of the data in Table 5.4 are derived from pot experiments, and protozoan numbers found in pot experiments generally exceed those from field experiments. The data represent the standing crops that are the net result of protozoan reproduction and simultaneous death by predation or lysis. The impact of protozoa in any habitat, including the rhizosphere, is determined more by their generation times than by the size of their standing crop. Unfortunately, calculations of the generation times cannot be made reliably in field soils at present. Differences in the limits of the rhizosphere chosen by different authors is another difficulty. In many cases, the soil adhering to the roots was regarded as rhizosphere soil. In some pot experiments, however, all the soil from the planted pots was regarded as rhizosphere soil. It seems likely in these latter cases that the protozoan numbers in the rhizosphere were underestimated. Notwithstanding all methodological problems, the ratio between protozoan biomass in the rhizosphere (R) and bulk soil (S) is a convenient indication of the rhizosphere-effect on protozoan numbers. The average R to S ratio calculated from the data in Table 5.4 is 6 with a high variation (SD = 7.1). It clearly demonstrates that favourable conditions for protozoan production exist in the rhizosphere. This R to S ratio is comparable with the ratio of 4 found between 'hot spots' of soil organic matter and bulk soil of agricultural fields in The Netherlands (Van Noordwijk *et al.*, 1993). The high R:S variation indicates that the positive effect of the rhizosphere on protozoa is not always very clear. For example, Griffiths (1990) generally found a positive effect of the rhizosphere on protozoa in pot experiments with four different plants (see Table 5.4), but only in the case of peas and fertilized barley was the effect statistically significant. Furthermore, the protozoan biomass in his experiments was an order of magnitude lower than the nematodal biomass and also lower than the protozoan biomass observed by Clarholm (1981) and De Ruiter *et al.* (1993) under winter wheat and Zwart *et al.* (unpublished results) under barley. Griffiths (1990) assumed that in his experiments the contribution of protozoa to the turnover of microbial biomass in the rhizosphere was small when compared to nematodes.

Very little is known about whether any protozoan species are specific to the rhizosphere. In contrast to a report of Biczók (1965), no protozoan genera were found by Darbyshire and Greaves (1967) to be specific for the rhizosphere of white mustard, white clover or perennial ryegrass, but these plant species appeared to stimulate the same flagellate species to varying degrees. These differences were probably the result of different degrees of microfloral production and root exudation with different plant species. Further studies of protozoan specificity for the rhizosphere of other plant species are required.

## Effects of Rhizosphere Protozoa on Plants

An increased mineralization and subsequent uptake by plants of nitrogen in the rhizosphere has been observed on several occasions. Haider *et al.* (1987) found that 500 mg organic N was mineralized and taken up by maize in glasshouse experiments, but the amount mineralized from soil in unplanted pots was only 150 mg. Paustian *et al.* (1989) calculated that 8.6 g m$^{-2}$ more N was mineralized when a soil was planted with a grass ley

**Table 5.4.** Protozoan populations in the rhizosphere (R) and bulk soil (S) of different plants. C, conditions; F, field; P, pot; U, units; O, organisms; P, protozoa; A, amoebae; F, flagellates; C, ciliates; na, no data available.

| Plant | C | R | S | U | R:S mean | R:S average | O | Reference |
|---|---|---|---|---|---|---|---|---|
| Barley | F | $48 \times 10^2$ | $14 \times 10^2$ | $g^{-1}$ | 3.4 | | P | Rouatt *et al.* (1960) |
| | F | $5 \times 10^4$ | $1 \times 10^4$ | $g^{-1}$ | 5 | | A | Clarholm (1989) |
| | P | 126 | 50 | ng $g^{-1}$ | 2.5 | 1.1–3.4 | P[a] | Griffiths (1990) |
| | P | 122 | 92 | ng $g^{-1}$ | 1.3 | 0.7–1.3 | P[b] | Griffiths (1990) |
| | P | 154 | 90 | ng $g^{-1}$ | 1.7 | 0.3–1.7 | P[c1] | Griffiths (1990) |
| | P | 1200 | 706 | ng $g^{-1}$ | 1.7 | 1.6–2.6 | P[c2] | Griffiths (1990) |
| | P | 315 | 168 | ng $g^{-1}$ | 1.9 | 1.6–2.9 | P[d1] | Griffiths (1990) |
| | P | 2165 | 1075 | ng $g^{-1}$ | 2.0 | 2.0–6.7 | P[d2] | Griffiths (1990) |
| | P | 210 | 98 | ng $g^{-1}$ | 2.1 | 1.3–2.6 | P | Griffiths (1990) |
| Wheat | F | na | na | | 2.4 | | P | Rouatt *et al.* (1960) |
| | P | na | na | | 2.2 | | A[e1] | Gamard *et al.* (1990) |
| | P | na | na | | 13.2 | | A[e2] | Gamard *et al.* (1990) |
| | P | 200 | 50 | µg $g^{-1}$ | 4 | | A[f] | Clarholm (1981) |
| Sugar beet | F | na | na | | 3.6 | | A | Kuikman (unpublished) |
| | F | na | na | | 5.2 | | A | Kuikman (unpublished) |
| | F | na | na | | 4.7 | | F | Kuikman (unpublished) |
| | F | na | na | | 8.7 | | F | Kuikman (unpublished) |
| Mangels | F | na | na | | na | 2.4–4.4 | P[g1] | Katznelson (1946) |
| | F | na | na | | na | 1.2–23 | P[g2] | Katznelson (1946) |

| | | | | | | | |
|---|---|---|---|---|---|---|---|
| Turnips | P | 95 | 56 | ng g⁻¹ → $ng\,g^{-1}$ | 1.7 | 0.9–2.4 | P[a] | Griffiths (1990) |
| Peas | P | 228 | 24 | $ng\,g^{-1}$ | 9.5 | 0.95–9.5 | P[a] | Griffiths (1990) |
| | P | 446 | 103 | $ng\,g^{-1}$ | 4.3 | | P[b] | Griffiths (1990) |
| *Lolium perenne* | P | 178 | 105 | $ng\,g^{-1}$ | 1.7 | 1.5–3.9 | P[a] | Griffiths (1990) |
| | P | $8.0 \times 10^4$ | $2.6 \times 10^4$ | $g^{-1}$ | 3.1 | 1.6–3.1 | A[h1] | Darbyshire and Greaves (1967) |
| | P | $5.7 \times 10^4$ | $1.4 \times 10^4$ | $g^{-1}$ | 4.1 | 1.3–4.1 | F[h1] | Darbyshire and Greaves (1967) |
| | P | $0.7 \times 10^4$ | $0.03 \times 10^4$ | $g^{-1}$ | 22 | 0–22 | C[h1] | Darbyshire and Greaves (1967) |
| *Trifolium repens* | P | $28.2 \times 10^4$ | $3.2 \times 10^4$ | $g^{-1}$ | 8.9 | 2.1–8.9 | A[h2] | Darbyshire and Greaves (1967) |
| | P | $12.2 \times 10^4$ | $3.3 \times 10^4$ | $g^{-1}$ | 3.4 | 2.9–6.5 | F[h2] | Darbyshire and Greaves (1967) |
| | P | 0 | $0.03 \times 10^4$ | $g^{-1}$ | 0 | 0–4 | C[h2] | Darbyshire and Greaves (1967) |
| *Sinapsis alba* | P | $18.1 \times 10^4$ | $2.4 \times 10^4$ | $g^{-1}$ | 5.2 | 2.8–12.5 | A[h3] | Darbyshire and Greaves (1967) |
| | P | $45 \times 10^4$ | $3.5 \times 10^4$ | $g^{-1}$ | 13 | 3.1–92 | F[h3] | Darbyshire and Greaves (1967) |
| | P | $1.8 \times 10^4$ | $0.05 \times 10^4$ | $g^{-1}$ | 35 | | C[h3] | Darbyshire and Greaves (1967) |

[a] Active cells after 10 weeks of growth in a clay-loam soil.
[b] Active cells after 3 weeks of growth in a sandy soil.
[c1] Total cells after 7 weeks of growth in a clay-loam soil without fertilizer.
[c2] Active cells after 7 weeks of growth in a clay-loam soil without fertilizer.
[d1] Total cells after 7 weeks of growth in a clay-loam soil with a plant nutrient solution applied.
[d2] Active cells after 7 weeks of growth in a clay-loam soil with a plant nutrient solution applied.
[e1] Soil inoculated with *Azospirillum*.
[e2] Soil not inoculated.
[f] During peak in biomass after rewetting of dry soil.
[g1] No fertilizer applied.
[g2] With farmyard manure.
[h1] After 7 days of growth.
[h2] After 72 days of growth.
[h3] After 29 days of growth.

than barley and all of this extra N was taken up by the plants. Lethbridge and Davidson (1983) and Texter and Billès (1990) also showed that microbial $^{15}N$ in the rhizosphere can be assimilated by wheat plants.

A number of reports of protozoa stimulating nutrient uptake by plants have been published. Elliot *et al.* (1979) found that blue grama grass (*Bouteloua gracilis*) took up significantly more N from fertilized soil in the presence of amoebae than in their absence. The size of this stimulatory effect depended on the initial N content of the soil. The increase in uptake of N induced by protozoa was 11.6, 232 and 11.5% at low (45 μg), medium (95 μg) and high (245 μg $NH_4^+$ N $g^{-1}$ soil) soil N content, respectively. The greatest effect did not occur at the lowest soil N level, as expected. Clarholm (1985, 1989) estimated, from the fluctuations in bacterial and protozoan numbers, that 10–17% of the N taken up by barley fertilized with normal levels of N could be accounted for by the bacteria–protozoa interactions. Ritz and Griffiths (1987) added glucose and nitrate to pots with a sandy soil and planted the pots with ryegrass. When no glucose was added, 75% of the nitrate was lost by leaching, whereas in the presence of glucose only 20% was lost by this process. They suggested that most of the nitrate in the nitrate-amended pots was assimilated by microorganisms. As a result of subsequent predation of microorganisms by protozoa, the ryegrass plants were able to assimilate 44% of this immobilized N. Kuikman *et al.* (1990) grew wheat in soil microcosms with bacteria alone or with bacteria and protozoa inoculated at different concentrations. In the presence of a protozoan inoculum diluted ten-fold, the wheat plants took up 9% more N than in the absence of protozoa. With an undiluted protozoan inoculum the N uptake was increased by 17% (Table 5.5). Clarholm (1985) and Kuikman and Van Veen (1989) showed that bacteria could mineralize N from soil organic matter (SOM) and subsequent protozoan grazing could release more of this N to plants. Using a simulation model of the detrital food web, Hunt *et al.* (1987) estimated that 14% of the N taken up by plants

**Table 5.5.** Numbers ($g^{-1}$ dry soil) of bacteria (B) and protozoa (P), total plant N, plant $^{15}N$, % $^{15}N$ in plant N and the ratio $^{14}C$-$CO_2$: $^{15}N$-plant (means of three replicated microcosms, Kuikman *et al.*, 1990). The protozoan inocula were diluted 1:10 (Bp) or undiluted (BP). B = pure bacteria treatment.

| Treatment | B ($10^8$) | P ($10^3$) | Plant N (mg) | Plant $^{15}N$ (mg) | % $^{15}N$ | $^{14}C$:$^{15}N$ (Bq mg$^{-1}$) |
|---|---|---|---|---|---|---|
| B | 1.30 | – | 47.95 | 0.577 | 1.161 | 17.8 |
| Bp | 0.76 | 2.77 | 52.24 | 0.313 | 1.173 | 22.0 |
| BP | 0.76 | 3.01 | 55.96 | 0.627 | 1.119 | 24.2 |
| LSD (*P*=0.05) | 0.47 | 1.75 | 4.19 | 0.059 | 0.042 | |

in a shortgrass prairie could be accounted for by amoebal predation of bacteria.

Soil protozoa can also affect enzymatic activity in the rhizosphere. Gould *et al.* (1979) studied the effect of bacteria and protozoa on rhizosphere phosphatase activity of the grass *Bouteloua gracilis*. They concluded that both bacteria (*Pseudomonas fluorescens*) and the combination of these bacteria and amoebae (*Acanthamoeba* sp.) stimulated acid phosphatase activity of the plants 2.3- and 2.6-fold, respectively. They also showed that the properties of this newly formed enzyme were similar to plant phosphatase and differed from acid phosphatase of the amoebae.

Detrimental effects of rhizosphere protozoa have also been reported. Ramirez and Alexander (1980) demonstrated that protozoa controlled the number of root-nodule bacteria around the roots of germinating seeds of *Phaseolus vulgaris*. Exudates from the roots initially stimulated the number of rhizobia, but subsequently protozoa fed on these bacteria and decreased rhizobial numbers. The final size of the *Rhizobium* population depended on the time of inoculation of bacteria and nutrient amendments. When *Rhizobium* was inoculated together with mannitol, the number of *Rhizobium* rapidly increased for 2 days and then decreased. When the rhizobia were inoculated 2 days after the mannitol amendment, the bacteria reproduced ten times more slowly. A delayed bacterial inoculation of 3 days completely suppressed the growth of *Rhizobium*. In contrast, suppression of the protozoa by thiram or a combination of Triton X-100 and cycloheximide resulted in an increase in *Rhizobium* numbers. Lennox and Alexander (1981) found that suppressing protozoa by the use of thiram, resulted in enhanced growth of *P. vulgaris* and greater initial numbers of root nodules formed by a thiram-resistant strain of *R. phaseoli*.

## Possible explanations of protozoan effects on plants

Protozoa and microbivorous nematodes are the most important predators of bacteria and fungi in terrestrial ecosystems (Bamforth, 1985; Ingham *et al.*, 1986; Zwart and Brussaard, 1991; De Ruiter *et al.*, 1993). The carbon, nitrogen and phosphorus contents of protozoa and their prey are very similar. Only 10–40% of the prey C is incorporated into newly formed protozoan biomass with most of the remainder being used in respiration. The excess N and P in the prey is excreted into the environment chiefly as inorganic forms (Meiklejohn, 1930, 1932; Stout, 1973; Zwart and Darbyshire, 1992) that can be assimilated daily by plants or other microorganisms. The liberation of this excess mineral N and P from microorganisms by protozoa is the commonest explanation for the increased uptake by plants in the presence of protozoa (Elliott *et al.*, 1979; Clarholm, 1985, 1989; Ritz and Griffiths, 1987; Gamard *et al.*, 1990; Kuikman *et al.*, 1990; Texter and

Billès, 1990). Other authors have suggested that protozoan grazing of bacteria also affects bacterial activity in aquatic systems (Johannes, 1965) and in soil (Coleman *et al.*, 1978; Stout, 1980; Pussard and Rouelle, 1986). The total metabolic activity of the bacterial population diminished by grazing can be as high or even higher than the ungrazed bacterial population (Kuikman *et al.*, 1990). The causes for this stimulation are uncertain. According to Monod kinetics, protozoan excretion of essential nutrients could stimulate bacterial activity under nutrient-stressed conditions. In addition, protozoa may excrete specific metabolic products or biocatalysts that stimulate microorganisms (Hervey and Greaves, 1941; Nikoljuk, 1969; Levrat, 1990). Recent investigations of this latter subject are discussed in Chapter 6.

Elliot *et al.* (1979) attempted to explain the observed increase in plant-available nitrogen in microcosm studies by suggesting that with the release of carbohydrates from plant roots, the rhizosphere microflora will assimilate N and other nutrients to satisfy their metabolic requirements. Such microbial immobilization of N will increase the diffusion gradients between the rhizosphere and other soil regions more remote from the roots. Under this interpretation, N and other nutrients should accumulate near the roots and the release of some of these nutrients after protozoan predation could benefit plants, providing the roots can compete with other rhizosphere microorganisms and sequestration by abiotic processes. These nutrient fluxes in the presence of protozoa have not been measured in experiments under a range of conditions and so this hypothesis has not been tested. Goring and Clark (1948), however, found that less mineral N accumulated in cropped soils than in fallow soils and microorganisms had a higher affinity for ammonium than plant roots. By contrast, Zak *et al.* (1990) found that plant roots of *Allium tricoccum* were better competitors for ammonium than heterotrophic and nitrifying bacteria. Verhagen (1992) obtained similar results with *Plantago lanceolata*. Rosswall (1982) reported Michaelis constants of 0.28–2.4, 0.2 and 8.0–18 mg N l$^{-1}$ for plants, heterotrophic and nitrifying bacteria, respectively.

Clarholm (1985) extended the root exudation and rhizosphere microbial interaction models of Rovira *et al.* (1979) (Fig. 5.3a) and Trofymow and Coleman (1982) (Fig. 5.3b) by suggesting that most carbonaceous compounds are released near the growing root-tip (Fig 5.3c). In her model, soil microorganisms could utilize these compounds as a source of carbon and energy to mineralize organic nitrogen associated with adjacent soil organic matter. The increased population of bacteria would attract protozoa, which would consume the bacteria and release ammonium N close to the root. In this way, plants would be able to utilize organic N that would be otherwise inaccessible. Texter and Billès (1990) proposed a similar explanation for the mineralization of soil organic N in the rhizosphere of wheat, but they did not attribute any role to protozoa or other soil fauna. Robinson *et al.* (1989)

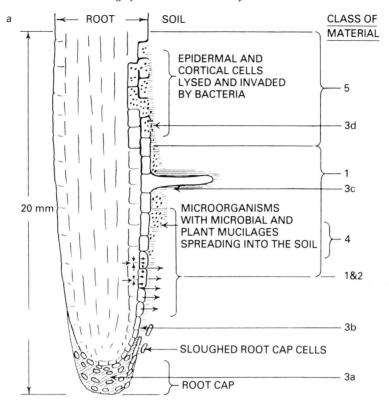

**Fig. 5.3a.** Models of interactions between roots, microorganisms and predators. Origins of organic matter in the rhizosphere: 1, exudates of low molecular weight released by non-metabolic processes; 2, actively secreted compounds; 3, plant mucilages; a, root cap material; b, sloughed root cap cells; c, mucilage secreted by epidermal cells, including root hairs; d, mucilage produced by bacterial degradation of dead epidermal cells; 4, mucigel, a product of the entire root–soil–microbial complex; 5, lysates from autolysis of older epidermal cells (Rovira *et al.*, 1979).

simulated the processes in root-induced mineralization and they concluded that at least part of the N taken up by plants originated from soil organic matter (SOM). Griffiths and Robinson (1992), however, after further simulations concluded that plant-derived C is relatively unimportant in inducing N mineralization from SOM. Clarholm (1985) estimated that the process described in her model could provide barley plants with 4–30 kg N from mineralization of SOM induced by root-derived carbon. Clarholm (1985) assumed that the rate of root exudation was between 7.5 and 55 kg ha$^{-1}$; 42% C and no N occurred in the exudate; all the N was derived from SOM; a bacterial growth efficiency of 0.4; a bacterial C:N ratio of 10; that

108

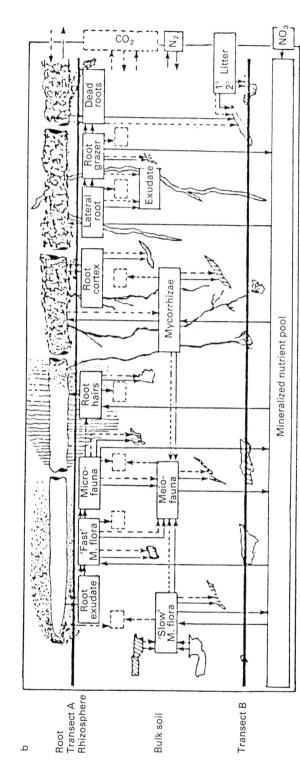

**Fig. 5.3b.** Models of interactions between roots, microorganisms and predators. Conceptual models of the root–rhizosphere–soil system (Coleman *et al.*, 1984).

c

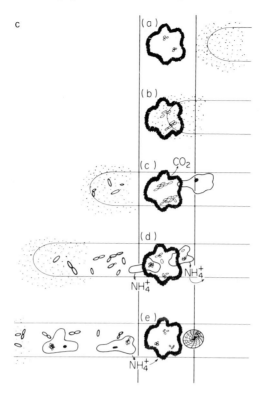

**Fig. 5.3c.** Models of interactions between roots, microorganisms and predators. Interaction of growing roots, soil organic matter and microbial activity resulting in the release of ammonium N from soil organic matter (Clarholm, 1985).

33% of the bacterial N is directly released as ammonium, together with 66% of the produced protozoan N; and a growth period of 60 days. Clarholm (1985) also assumed that all C exuded by the roots was assimilated by microorganisms and used to mineralize organic N from the soil. With these assumptions, she calculated for barley that an extra 25 kg N at the most may be released per hectare from SOM. An important prerequisite to Clarholm's calculations is that sufficient decomposable N is available in the rhizosphere. The estimated amount of decomposable organic N, rhizosphere volumes and root lengths are presented in Table 5.6. Van Noordwijk and Brouwer (1991) have reviewed data on root lengths of agricultural plants (Table 5.7). The total root length in barley is $0.8 \times 10^8$ m ha$^{-1}$, i.e. sufficient to provide a rhizosphere volume large enough to release 25 kg inorganic N if 2% of the SOM is decomposed. All plants listed in Table 5.7 have roots long enough for that purpose, assuming root exudation at the level found

**Table 5.6.** Estimated volumes of rhizosphere soil and root lengths needed to provide 25 kg mineral N ha[-1] from 500 kg soil organic matter (SOM) at different percentages of decomposition, assuming a soil bulk density of 1.3, SOM content of rhizosphere soil of 2.5% (50% C and 5% N), root diameter 0.2 mm and rhizosphere diameter 2 mm.

| SOM decomposition (%) | Volume of rhizosphere soil (m$^3$) | Root length (m ha$^{-1}$ × 10$^8$) |
|---|---|---|
| 2 | 770 | 0.56 |
| 10 | 154 | 0.11 |
| 25 | 62 | 0.045 |

**Table 5.7.** Total root length (m ha$^{-1}$ × 10$^8$) of different agricultural crops (adapted from Van Noordwijk and Brouwer, 1991).

| Plant | Root length |
|---|---|
| Wheat | 2.5 |
| Barley | 0.8 |
| Oats | 0.3 |
| Maize | 0.9 |
| Sorghum | 3.0 |
| Grass | 6.3 |
| Soyabean | 0.8 |

for barley. It is concluded that if root exudates contain no N at all, theoretically, all exuded C could be used to mobilize N from SOM.

As protozoa excrete mineral N during the assimilation of bacteria and fungi, the total amount of N liberated from a microbial biomass by protozoan predation will depend on the protozoan population dynamics. In this section we are concerned with N excretion by rhizosphere protozoa in relation to their population dynamics. The calculations are based upon the assumptions that are listed in Table 5.8. In addition we have assumed that the rhizosphere volume is constant during the whole growth period, which will obviously overestimate N excretion, that N is not a limiting nutrient for bacterial growth in the rhizosphere and that the total protozoan biomass in the rhizosphere will remain constant. Based on these assumptions, the total protozoan production, the concomitant consumption of bacteria and the protozoan N excretion can be calculated (Table 5.9). With these assumptions, an extra 25 kg N cannot be mineralized by protozoa if the standing population biomass is of the order of magnitude as described by Griffiths and Robinson (1992). Instead, the amount of mineralized N does not exceed 1.1 kg ha$^{-1}$ under the most beneficial conditions, i.e. a high protozoan growth rate of 0.16 d$^{-1}$ and a low bacterial C:N ratio of 3. Griffiths (1990)

**Table 5.8.** Assumptions for the calculations on the N excretion by protozoan grazing in the rhizosphere (see also Table 5.9).

| | |
|---|---|
| 1. Rhizosphere surface area | $13.8 \times 10^6$ m$^2$ ha$^{-1}$ |
| 2. Root length | |
|     Barley | $0.8 \times 10^8$ m ha$^{-1}$ |
|     Wheat | $2.5 \times 10^8$ m ha$^{-1}$ |
| 3. Total growth period | 60 days |
| 4. Yield of protozoa | 0.4 |
|     (kg C produced/kg C consumed) | |
| 5. Yield of bacteria | 0.3 |
| 6. C:N of bacteria[a] | 3, 5 or 10 |
|     C:N of protozoa | 5 |
| 7. Fraction of protozoa produced which becomes available as mineral N[b] | 0.66 |
| 8. R:S for protozoa[c] | 6 |
| 9. Turnover rate of protozoa[d] | 0.02, 0.07 or 0.16 d$^{-1}$ |
| 10. Protozoan biomass in rhizosphere soil[e] | |
|     Griffiths (1990) | 0.1–0.32 μg C g$^{-1}$ soil |
|     Clarholm (1981) | 200 μg C g$^{-1}$ soil |
|     Andrén *et al.* (1990) | 116 μg C g$^{-1}$ soil |
|     De Ruiter *et al.* (1993) | 38.4–69.8 μg C g$^{-1}$ soil |

[a] The C:N ratio of bacteria and fungi varies with the cultivation conditions. For bacteria, values of 3.5–4.4 (Jenkinson, 1976), 2.8–22.7 (Elliott *et al.*, 1983), 3.7 (Bakken, 1985; Lee and Furman, 1987), 3.8–6.7 (Van Veen and Paul, 1979), 5.6 (Anderson and Domsch, 1980), 3–5 (Paul and Clark, 1989) and 4.4–17.2 (Tezuka, 1990) have been reported. For ammonium- and for glucose-limited cultures of a mixture of a heterotrophic and two nitrifying bacteria (Verhagen and Laanbroek, 1991) values of 3.5–3.6 have been reported.
[b] Clarholm (1985).
[c] See page 101.
[d] The figure of 0.16 d$^{-1}$ is used by Hunt *et al.* (1987) and by De Ruiter *et al.* (1993). Fenchel and Finlay (1983) and Heal (1971) give specific respiration rates for starved protozoa from which a turnover rate of 0.02 and 0.07 day$^{-1}$ can be calculated for naked amoebae and flagellates, respectively.
[e] For the conversion from biomass g$^{-1}$ soil to biomass C m$^{-3}$, a soil bulk density of 1.3 and a C content of 50% was assumed.

also concluded from his experiments that the contribution of protozoa to plant N supply was small. If the protozoan standing population biomass is of the order of magnitude found by Clarholm (1981, 1985), then 25 kg extra N can be mineralized by protozoa under all conditions, except when the protozoan growth rate is low and the bacterial C:N ratio is high. An intermediate situation occurs when the standing population of protozoa is in between these two estimates. The correct size of the protozoan biomass in the rhizosphere (Table 5.4) is a crucial factor determining the validity of the final conclusions about protozoan mineralization of N. Another important factor in the estimations is the mean generation time of the protozoan population. Hunt *et al.* (1987) tested the validity of their

**Table 5.9.** Protozoan production, bacterial consumption and nitrogen excretion by protozoa in the rhizosphere of barley (in kg ha⁻¹) based upon various standing populations in the rhizosphere, different growth rates of protozoa and different C:N ratios of bacteria.

| | Growth rate d⁻¹ | Protozoan biomass ($\mu$g C g⁻¹ soil) | | | | | |
|---|---|---|---|---|---|---|---|
| | | 0.1[a] | 0.32[a] | 200[b] | 116[c] | 38.4[d] | 69.8[d] |
| Protozoan production | 0.02 | 0.09 | 0.26 | 172.2 | 99.9 | 33.1 | 60.1 |
| | 0.07 | 0.30 | 0.96 | 602.8 | 349.6 | 115.7 | 210.4 |
| | 0.16 | 0.69 | 2.2 | 1377.8 | 799.1 | 264.5 | 480.8 |
| Bacterial consumption | 0.02 | 0.21 | 0.69 | 430.6 | 249.7 | 82.7 | 150.3 |
| | 0.07 | 0.75 | 2.41 | 1507.0 | 874.0 | 289.3 | 525.9 |
| | 0.16 | 2.15 | 6.90 | 4305.6 | 2497.0 | 826.7 | 1502.7 |
| N excretion | 0.02 | 0.04 | 0.14 | 86.1 | 49.9 | 16.5 | 30.1 |
| (C:N bacteria = 3) | 0.07 | 0.15 | 0.48 | 301.4 | 174.8 | 57.9 | 105.2 |
| | 0.16 | 0.34 | 1.10 | 688.9 | 399.5 | 132.6 | 240.4 |
| (C:N bacteria = 5) | 0.02 | 0.03 | 0.08 | 51.6 | 30.0 | 9.9 | 18.0 |
| | 0.07 | 0.09 | 0.29 | 180.8 | 104.9 | 34.7 | 63.1 |
| | 0.16 | 0.21 | 0.66 | 413.3 | 239.7 | 79.4 | 144.3 |
| (C:N bacteria = 10) | 0.02 | 0.01 | 0.04 | 25.8 | 15.0 | 5.0 | 9.0 |
| | 0.07 | 0.05 | 0.15 | 90.4 | 52.4 | 17.4 | 31.5 |
| | 0.16 | 0.1 | 0.33 | 206.7 | 119.9 | 39.7 | 72.5 |

[a] Griffiths (1990).
[b] Clarholm (1981).
[c] Andrén *et al.* (1990).
[d] De Ruiter *et al.* (1993).

model calculations of N mineralization by soil organisms. They demonstrated that the generation times of protozoa are directly related to their impact on mineralization: changing the turnover rate by a factor of 2 will also change the protozoan mineralization by a factor of 2. This is confirmed by the calculations in Table 5.9. Stout and Heal (1967) estimated a possible turnover rate for small amoebae and flagellates of 80–300 times per year (0.22–0.82 d⁻¹) from daily counts of soil protozoa by Cutler *et al.* (1992). Similarly, from respiration rates of starved protozoan cells (Heal, 1971; Fenchel and Finlay, 1983), a turnover rate of 0.02 d⁻¹ and 0.07 d⁻¹ can be estimated for amoebae and heterotrophic flagellates. These latter estimates of turnover rate, that are also used by Hunt *et al.* (1987) in computer simulations, are only sufficient to account for maintenance requirements. A death rate based upon both maintenance requirements and predation of protozoa by other organisms (protozoa, nematodes, microarthropods, earthworms, etc.) would seem to be more realistic. However, we have been unable to find any data in the literature on the rate at which protozoa are consumed by other soil animals. Perhaps adding isotope-labelled protozoa to soil and following isotope decay rates might give some indication of this process in the future. Finally, variations in soil temperature and moisture were not included in the calculations and may affect the outcome. It was assumed that the total protozoan biomass was continuously active during the growth period of 60 days. As many protozoa are often in an inactive encysted state in soil, due to unfavourable abiotic conditions, the values in Table 5.9 are overestimates.

In her calculations, Clarholm (1985) assumed that the root exudates contained no nitrogen. In reality, most plants exude some organic N and most of this can be readily utilized by soil microorganisms. Only when the C:N ratio of root exudates is very high, i.e. above 15, with a bacterial C:N ratio of 5 and a carbon yield of 0.3, would soil-derived N be required for the formation of new microbial biomass. At lower exudate C:N ratios and/ or at higher bacterial C:N ratios, at least part of the exudate N is directly mineralized by bacteria. Also, when C and N in the exudates are separated either in space or time, soil microorganisms may be more likely to utilize SOM N as the major N source, especially when the C:N ratio in the exudates is large. More precise information about when and where N exudation occurs in the rhizosphere should improve these calculations. Although Robinson *et al.* (1989) cite differences in the C:N ratio of root exudates from the literature, there is little information about the variation in the quantity and quality of root exudates with time and space.

It seems probable that Clarholm (1985) overestimated the quantity of SOM nitrogen released by root exudation. Although the amount of organic N available in the rhizosphere is sufficiently large and the protozoan biomass turnover rate in many situations large enough to release 25 kg of N ha$^{-1}$ during a growing season, much of the N probably originated from the plant itself and not from SOM. This does not mean that grazing by soil fauna in general, and protozoa in particular, is of no consequence at all for the uptake of N by plants. The part of the organic N exuded by plants that is subsequently assimilated by microorganisms will only be released again as a result of predation by protozoa and other soil animals. The rate at which it may become available for plants again largely depends on the feeding rate of these bacterial grazers. We have modified the flows of N of the model of Clarholm (1985) as illustrated in Fig. 5.4

## Rhizosphere Protozoa and Plant Diseases

Soil protozoa have the potential to suppress diseases caused by pathogenic bacteria and fungi, because they can consume some of these pathogens. This

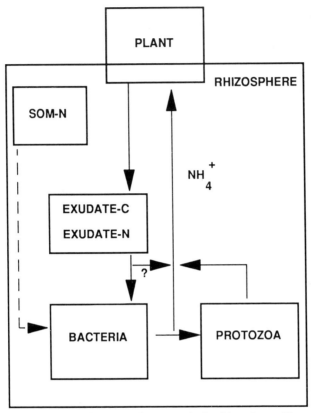

**Fig. 5.4.** Conceptual model of the role of protozoa in the rhizosphere with respect to flows of carbon and nitrogen. The dashed arrow indicates a flow of N that may occur when the C:N ratio of exudates is large. The question mark indicates a flow of N from exudates to the plant that may occur during the metabolism of plant exudates by microorganisms when the C:N ratio of exudates is small. SOM-N = Soil organic matter nitrogen.

also may be the case for the few examples of nematophagous amoebae that have been described (Weber *et al.*, 1952; Zwillenberg, 1952; Old and Darbyshire, 1978). To our knowledge the suppression of plant pathogenic bacteria by protozoa has only been described by Habte and Alexander (1975). In contrast, the predation of plant pathogenic fungi by soil amoebae has been observed on several occasions. The subject has been reviewed by Old and Chakraborty (1986) and by Curl (1988). It is also discussed in more detail in Chapter 6. Obligate mycophagous ciliates (Petz *et al.*, 1985) and a flagellate (Hekman *et al.*, 1992) that digest pathogenic fungi have also been described.

The impact of protozoa on suppressing plant pathogens may be greater in the rhizosphere than in the bulk soil, because of the higher protozoan concentration in the rhizosphere. Unfortunately, many plant pathogens profit from plant exudates (Newman, 1978) so it is likely that their populations will be larger in the rhizosphere. It is also known that protozoa in soil never eliminate their prey completely (Danso *et al.*, 1975; Habte and Alexander, 1975; Alabouvette *et al.*, 1981). Hekman *et al.* (1992) isolated an obligate mycophagous flagellate that reduced the number of conidia of the fungus *Drechmeria coniospora* in soil by only 20%. In addition, most plant pathogens have higher growth rates than their protozoan predators so it seems very unlikely that protozoa will suppress plant pathogens completely in soil. If there is an increase in the number of crops grown in liquid media in the future, mycophagous protozoa may become important biological control agents. For example, the flagellate described by Hekman *et al.* (1992) almost completely eliminated fungal spores from liquid media.

## Concluding Remarks

The rhizosphere is a suitable environment for soil protozoa and in this region protozoa may have beneficial effects on plants. The dimensions of the carbon and nitrogen flows to and from plants via soil protozoa are difficult to estimate as long as some fundamental protozoan problems remain unresolved. These are the enumeration of the active protozoan biomass in soil, the turnover rates of their populations and the chemical composition of these organisms and their food in the rhizosphere soil. The solution of such problems should be major objectives of future research.

## Acknowledgement

We thank John Darbyshire for his encouragement and stimulating advice on this chapter.

## References

Alabouvette, C., Lemaitre, I. and Pussard, M. (1981) Densité de population de l'amibe mycophage *Thecameoba granifera* s. sp *minor* (Amoebida, Protozoa). Mesure et variations expérimentales dans le sol. *Revue d'Écologie et de Biologie du Sol* 18, 179–192.

Anderson, J.P.E. and Domsch, K.H. (1980) Quantities of plant nutrients in the microbial biomass of selected soils. *Soil Science* 130, 211–216.

Andrén, O., Lindberg, T., Boström, U., Clarholm, M., Hansson, A.-C., Johansson, G., Lagerlöf, J., Paustian, K., Persson, J., Pettersson, R., Schnürer, J., Sohlenius, B. and Wivstad, M. (1990) Ecology of arable land 5. Organic carbon and nitrogen flows. *Ecological Bulletins (Stockholm)* 40, 85–126.

Bakken, L.R. (1985) Separation and purification of soil bacteria from soil. *Applied and Environmental Microbiology* 49, 1482–1487.

Bamforth, S.S. (1985) The role of protozoa in litters and soil. *Journal of Protozoology* 32, 404–409.

Barber, D.A. and Lynch, J.M. (1977) Microbial growth in the rhizosphere. *Soil Biology and Biochemistry* 9, 305–308.

Beck, S.M. and Gilmour, C.M. (1983) Role of wheat exudates in associative nitrogen fixation. *Soil Biology and Biochemistry* 15, 33–38.

Biczók, F. (1965) Protozoa in the rhizosphere. In: *Progress in Protozoology*, 2nd International Conference of Protozoology, London 1965. Excerpta Medica Foundation, Amsterdam, p. 120.

Biondini, M., Klein, D.A. and Redente, E.F. (1988) Carbon and nitrogen losses through root exudation by *Agropyron cristatum, A. smithii* and *Bouteloua gracilis*. *Soil Biology and Biochemistry* 20, 477–482.

Boulter, D., Jeremy, J.J. and Wilding, M. (1966) Amino acids liberated into the culture medium by pea seedling roots. *Plant and Soil* 24, 121–127.

Cheshire, M.V. and Mundie, C. (1990) Organic matter contributed to soil by plant roots during the growth and decomposition of maize. *Plant and Soil* 121, 107–114.

Christiansen-Weniger, C., Groneman, A.F. and Van Veen, J.A. (1992) Associative $N_2$ fixation and root exudation of organic acids from wheat cultivars of different aluminium tolerance. *Plant and Soil* 139, 167–174.

Clarholm, M. (1981) Protozoan grazing of bacteria in soil. Impact and importance. *Microbial Ecology* 7, 343–350.

Clarholm, M. (1985) Possible roles for roots, bacteria, protozoa and fungi in supplying nitrogen to plants. In: Fitter, A.H., Atkinson, D., Read, D.J. and Usher, M.B. (eds), *Ecological Interactions in Soil*. Blackwell Scientific Publications, Oxford, pp. 355–365.

Clarholm, M. (1989) Effects of plant–bacterial–amoebal interactions on plant uptake of nitrogen under field conditions. *Biology and Fertility of Soils* 8, 373–378.

Coleman, D.C., Anderson, R.V., Cole, C.V., Elliott, E.T., Woods, L. and Campion, M.K. (1978) Trophic interactions in soils as they affect energy and nutrient dynamics. IV. Flows of metabolic and biomass carbon. *Microbial Ecology* 4, 373–380.

Coleman, D.C., Ingham, R.E., McLellan, J.F. and Trofymow, J.A. (1984) Soil nutrient transformations in the rhizosphere via animal–microbial interactions. In: Anderson, J.M., Rayner, A.D.M. and Walton, D.W.H. (eds), *Invertebrate Microbial Interactions*. Cambridge University Press, Cambridge, pp. 35–58.

Coûteaux, M.-M., Faurie, G., Palka, L. and Steinberg, C. (1988) Le relation prédateur–proie (protozoaires–bactéries) dans les sols: rôle dans la régulation des populations et conséquences sur les cycles du carbone et de l'azote. *Revue d'Écologie et de Biologie du Sol* 25, 1–31.

Curl, E.A. (1988) The role of soil microfauna in plant disease suppression. *CRC Critical Reviews in Plant Sciences* 7, 175–196.

Cutler, D.W., Crump, L.M. and Sandon, H. (1922) A quantitative investigation of the bacterial and protozoan population of the soil, with an account of the protozoan fauna. *Philosophical Transactions of the Royal Society Series B* 211, 317–350.

Danso, S.K.A., Keya, S.O. and Alexander, M. (1975) Protozoa and the decline of *Rhizobium* populations added to soil. *Canadian Journal of Microbiology* 21, 884–895.

Darbyshire, J.F. and Greaves, M.P. (1967) Protozoa and bacteria in the rhizosphere of *Sinapis alba* L., *Trifolium repens* L. and *Lolium perenne* L. *Canadian Journal of Microbiology* 13, 1057–1068.

Darbyshire, J.F. and Greaves, M.P. (1973) Bacteria and protozoa in the rhizosphere. *Pesticide Science* 4, 349–360.

Darrah, P.R. (1991a) Models of the rhizosphere I. Microbial population dynamics around a root releasing soluble and insoluble carbon. *Plant and Soil* 133, 187–199.

Darrah, P.R. (1991b) Models of the rhizosphere II. A quasi three-dimensional simulation of the microbial population dynamics around a growing root releasing soluble exudates. *Plant and Soil* 138, 147–158.

De Ruiter, P.C., Moore, J.C., Zwart, K.B., Bouwman, L.A., Hassink, J., Bloem, J., De Vos, J.A., Marinissen, J.C.Y., Didden, W.A.M., Lebbink, G. and Brussaard, L. (1993) Simulation of nitrogen mineralization in the below-ground food webs of two winter wheat fields. *Journal of Applied Ecology* 30, 95–106.

De Willigen, P. and Van Noordwijk, M. (1987) Roots, plant production and nutrient use efficiency. PhD thesis, Agricultural University of Wageningen, The Netherlands.

Elliott, E.T., Coleman, D.C. and Cole, C.V. (1979) The influence of amoebae on the uptake of nitrogen by plants in a gnotobiotic soil. In: Harley, J.L. and Scott Russell, R. (eds), *The Soil–Root Interface*. Academic Press, London, pp. 221–229.

Elliott, L.F., Cole, C.V., Fairbanks, B.C., Woods, L.E., Bryant, R.J. and Coleman, D.C. (1983) Short-term bacterial growth, nutrient uptake, and ATP turnover in sterilised, inoculated and C-amended soil: the influence of N availability. *Soil Biology and Biochemistry* 15, 85–92.

Elliott, L.F., Gilmour, C.M., Lynch, J.M. and Tittemore, D. (1984) Bacterial colonization of plant roots. In: Todd, R.L. and Giddens, J.E. (eds), *Microbial–Plant Interactions*. American Society of Agronomy, Madison, pp. 1–16.

Fenchel, T. and Finlay, B.J. (1983) Respiration rates in hetrotrophic free-living protozoa. *Microbial Ecology* 9, 99–122.

Fogel, R. and Hunt, G. (1983) Contribution of mycorrhizae and soil fungi to nutrient cycling in a Douglas-fir ecosystem. *Canadian Journal of Forestry Research* 12, 219–232.

Foissner, W. (1987) Soil protozoa: Fundamental problems, ecological significance, adaptations in Ciliates and Testaceans, bioindicators, and guide to the literature. *Progress in Protistology* 2, 69–212.

Foster, R.C. and Dormaar, J.F. (1991) Bacteria-grazing amoebae *in situ* in the rhizosphere. *Biology and Fertility of Soils* 11, 83–87.

Gamard, P., Steinberg, C. and Faurie, G. (1990) Behaviour of bacterial populations in response to amoebal predation in the rhizosphere. *Symbiosis* 9, 367–376.

Goring, C.A.J. and Clark, F.E. (1948) Influence of crop growth on mineralization of nitrogen in soil. *Soil Science Society of America Proceedings* 13, 261–266.

Gould, W.D., Coleman, D.C. and Rubink, A.J. (1979) Effect of bacteria and amoebae on rhizosphere phosphatase activity. *Applied and Environmental Microbiology* 37, 943–946.

Griffiths, B.S. (1990) A comparison of microbial-feeding nematodes and protozoa in the rhizosphere of different plants. *Biology and Fertility of Soils* 9, 83–88.

Griffiths, B.S. and Robinson, D. (1992) Root-induced nitrogen mineralization: A nitrogen balance model. *Plant and Soil* 139, 253–263.

Habte, M. and Alexander, M. (1975) Protozoa as agents responsible for the decline of *Xanthomonas campestris* in soil. *Applied Microbiology* 29, 159–164.

Haider, K., Mosier, A. Heinemeyer, O. (1987) The effect of growing plants on denitrification at high soil nitrate concentrations. *Soil Science Society of America Journal* 51, 97–102.

Heal, O.W. (1971) Protozoa. In: Phillipson, J. (ed.), *Methods of Study in Quantitative Soil Ecology: Population, Production and Energy Flow.* Blackwell Scientific Publications, Oxford, pp. 51–71.

Hekman, W.E., Van den Boogert, P.J.H.F. and Zwart, K.B. (1992) The physiology and ecology of a novel, obligate mycophagous flagellate. *FEMS Microbiology Ecology* 86, 255–265.

Helal, H.M. and Sauerbeck, D. (1986) Effect of plant roots on carbon metabolism of soil microbial biomass. *Zeitschrift für Pflanzenernährung und Bodenkunde* 149, 181–188.

Hervey, R.J. and Greaves, J.E. (1941) Nitrogen fixation by *Azotobacter chroococcum* in the presence of soil protozoa. *Soil Science* 51, 85–100.

Hiltner, L. (1904) Über neuerer Erfahrungen und Probleme auf dem Gebiet der Bodenbakteriologie und unter besonderer Berücksichtigung der Gründungung und Brache. *Arbeitungen der Deutscher Landwirtschaftlichen Gesellschaft* 98, 59–78.

Hunt, H.W., Coleman, D.C., Ingham, E.R., Ingham, R.E., Elliot, E.T., Moore, J.C., Rose, S.L., Reid, C.P.P. and Morley, C.R. (1987) The detrital food web in a shortgrass prairie. *Biology and Fertility of Soils* 3, 57–68.

Ingham, E.R., Trofymow, J.A., Ames, R.N., Hunt, H.W., Morley, C.R., Moore, J.C. and Coleman, D.C. (1986) Trophic interactions and nitrogen cycling in a semi-arid grassland soil. I. Seasonal dynamics of the natural populations, their interactions and effects on nitrogen cycling. *Journal of Applied Ecology* 23, 615–630.

Janzen, H.H. (1990) Deposition of nitrogen in the rhizosphere by wheat roots. *Soil Biology and Biochemistry* 22, 1155–1160.

Jenkinson, D.S. (1976) The effects of biocidal treatment on metabolism in soil IV. The decomposition of fumigated organisms in soil. *Soil Biology and Biochemistry* 8, 203–208.

Johannes, R.E. (1965) Influence of marine protozoa on nutrient regeneration. *Limnology and Oceanography* 10, 434–442.

Johnen, B.G. and Sauerbeck, D.R. (1977) A tracer technique for measuring growth, mass and microbial breakdown on plant root during vegetation. In: Lohm, V. and Persson, T. (eds), *Soil Organisms as Components of Ecosystems.* Proceedings VIth International Zoology Colloquium. *Ecological Bulletins (Stockholm)* 25, 366–373.

Katznelson, H. (1946) The 'rhizosphere effect' of mangels on certain groups of soil microorganisms. *Soil Science* 62, 343–354.

Keith, H., Oades, J.M. and Martin, J.K. (1986) Input of carbon to soil from wheat plants. *Soil Biology and Biochemistry* 18, 445–449.

Klein, D.A., Frederick, B.A., Biondini, M. and Trlica, M.J. (1988) Rhizosphere microorganism effects on soluble amino acids, sugars and organic acids in the root zone of *Agropyron cristatum, A. smithii,* and *Bouteloua gracilis. Plant and Soil* 110, 19–25.

Kuikman, P.J. and Van Veen, J.A. (1989) The impact of protozoa on the availability of bacterial nitrogen to plants. *Biology and Fertility of Soils* 8, 13–18.

Kuikman, P.J., Jansen, A.G., Van Veen, J.A. and Zehnder, A.J.B. (1990) Protozoan predation and the turnover of soil organic carbon and nitrogen in the presence of plants. *Biology and Fertility of Soils* 10, 22–28.

Lee, S. and Furman, J.A. (1987) Relationships between biovolume and biomass of naturally derived marine bacterioplankton. *Applied and Environmental Microbiology* 53, 1298–1303.

Lennox, L.B. and Alexander, M. (1981) Fungicide enhancement of nitrogen fixation and colonization of *Phaseolus vulgaris* by *Rhizobium phaseoli. Applied and Environmental Microbiology* 41, 404–411.

Lethbridge, G. and Davidson, M.S. (1983) Microbial biomass as a source of nitrogen for cereals. *Soil Biology and Biochemistry* 15, 375–376.

Levrat, P. (1990) Contribution à l'étude des interactions entre protozoaires et microflore du sol: effet d'une amibe bactériophage *Acanthamoeba castellanii* sur la métabolisme de *Pseudomonas* fluorescents. PhD thesis, University of Claude Bernard, Lyon.

Martens, R. (1990) Contribution of rhizodeposition to the maintenance and growth of soil microbial biomass. *Soil Biology and Biochemistry* 23, 141–147.

Matsumoto, H., Okada, K. and Takahasi, E. (1979) Excretion products of maize roots from seedling to seed development stage. *Plant and Soil* 53, 17–26.

Meiklejohn, J. (1930). The relation between the numbers of a soil bacterium and the ammonia produced by it in peptone solutions; with some reference to the effect on this process of the prescence of amoebae. *Annals of Applied Biology* 17, 614–637.

Meiklejohn, J. (1932) The effect of *Colpidium* on the ammonia production by soil bacteria. *Annals of Applied Biology* 19, 584–608.

Merckx, R., Dijkstra, A., Den Hartog, A. and Van Veen, J.A. (1987) Production of root-derived material and associated microbial growth in soil at different nutrient levels. *Biology and Fertility of Soils* 5, 126–132.

Newman, E.I. (1978) Root microorganisms: their significance in the ecosystem. *Biological Reviews* 53, 511–554.

Newman, E.I. and Watson, A. (1977) Microbial abundance in the rhizosphere: a computer model. *Plant and Soil* 48, 17–56.

Nikoljuk, V.F. (1969) Some aspects of the study of soil protozoa. *Acta Protozoologica* 7, 99–109.

Nye, P.H. and Tinker, P.B. (eds) (1977) *Solute Movement in the Soil–Root System.* Studies in Ecology Volume 4. Blackwell Scientific Publications, Oxford.

Old, K.M. and Chakraborty, S. (1986) Mycophagous soil amoebae: their biology and significance in the ecology of soil-borne plant pathogens. *Progress in Protistology* 1, 163–194.

Old, K.M. and Darbyshire, J.F. (1978) Soil fungi as food for giant amoebae. *Soil Biology and Biochemistry* 10, 93–100.

Paul, E.A. and Clark, F.E. (1989) *Soil Microbiology and Biochemistry.* Academic Press, San Diego.

Paustian, K., Andrén, O., Clarholm, M., Hansson, A.-C., Johansson, G., Lagerlöf, J., Lindberg, T., Pettersson, R. and Sohlenius, B. (1989) Carbon and nitrogen budgets for four agro-ecosystems with annual and perennial crops, with and without N-fertilization. *Journal of Applied Ecology* 27, 60–84.

Petz, W., Foissner, W. and Adam, H. (1985) Culture, food selection and growth rate in the mycophagous ciliate *Grossglockneria acuta* Foissner 1980: first evidence of autochthonous soil ciliates. *Soil Biology and Biochemistry* 17, 871–875.

Pussard, M. and Rouelle, J. (1986) Prédation de la microflore. Effet des protozoaires sur la dynamique de population bactérienne. *Protistologica* 22, 105–110.

Ramirez, C. and Alexander, M. (1980) Evidence suggesting protozoan predation on

*Rhizobium* associated with germinating seeds and in the rhizosphere of beans (*Phaseolus vulgaris* L.) *Applied and Environmental Microbiology* 40, 492–499.

Ritz, K. and Griffiths, B.S. (1987) Effects of carbon and nitrate additions to soil upon leaching of nitrate, microbial predators and nitrogen uptake by plants. *Plant and Soil* 102, 229–237.

Robinson, D., Griffiths, B., Ritz, K. and Wheatley, R. (1989) Root-induced nitrogen mineralisation: A theoretical analysis. *Plant and Soil* 117, 185–193.

Römheld, V. (1990) The soil–root interface in relation to mineral nutrition. *Symbiosis* 9, 19–27.

Rosswall, T. (1982) Microbiological regulation of the biogeochemical nitrogen cycle. *Plant and Soil* 67, 15–34.

Rouatt, J.W., Katznelson, H. and Payne, T.B.M. (1960) Statistical evaluation of the rhizosphere effect. *Proceedings of the Soil Science Society of America* 24, 271–273.

Rovira, A.D. (1969) Plant root exudates. *The Botanical Review* 35, 35–37.

Rovira, A.D., Foster, R.C. and Martin, J.K. (1979) Note on terminology: origin, nature and nomenclature of the organic materials in the rhizosphere. In: Harley, J.L. and Scott Russell, R. (eds), *The Soil–Root Interface*. Academic Press, London, pp. 1–4.

Stout, J.D. (1973) The relationship between protozoan populations and biological activity in soils. *American Zoologist* 13, 193–201.

Stout, J.D. (1980) The role of protozoa in nutrient cycling and energy flow. *Advances in Microbial Ecology* 4, 1–50.

Stout, J.D. and Heal, O.W. (1967) Protozoa. In: Burges, A. and Raw, F. (eds), *Soil Biology*. Academic Press, London, pp. 149–195.

Texter, M. and Billès, G. (1990) The role of the rhizosphere on C and on N cycles in a plant–soil system. *Symbiosis* 9, 117–123.

Tezuka, Y. (1990) Bacterial regeneration of ammonium and phosphate as affected by the carbon:nitrogen:phosphorus ratio of organic substrates. *Microbial Ecology* 19, 227–238.

Trofymow, J.A. and Coleman, D.C. (1982) The role of bacterivorous and fungivorous nematodes in cellulose and chitin decomposition. In: Freckman, D.W. (ed.), *Nematodes in Soil Ecosystems*. University of Texas Press, Austin, pp. 117–138.

Trofymow, J.A., Coleman, D.C. and Cambarella, C. (1987) Rates of rhizodeposition and ammonia depletion in the rhizosphere of axenic oat roots. *Plant and Soil* 97, 333–344.

Vancura, V. (1988) Plant metabolites in soil. In: Vancura, V. and Kunc, F. (eds), *Soil Microbial Associations. Control of Structures and Functions*. Elsevier, Amsterdam, pp. 57–144.

Van Noordwijk, M. and Brouwer, G. (1991) Review of quantitative root length in agriculture. In: McMichael, B.L. and Persson, H. (eds), *Plant Roots and their Environment*. Elsevier, Amsterdam, pp. 515–525.

Van Noordwijk, M., De Ruiter, P.C., Zwart, K.B., Bloem, J., Moore, J.C., Van Faassen, H.G. and Burgers, S.L.G.E. (1993) Synlocation of biological activity, roots, cracks and recent organic inputs in a sugar beet field. *Geoderma* 56, 265–276.

Van Veen, J.A. and Paul, E.A. (1979) Conversion of biovolume measurements of soil organisms grown under various moisture tensions, to biomass and their nutrient content. *Applied and Environmental Microbiology* 37, 686–692.

Verhagen, F.J.M. (1992) Nitrification versus immobilization of ammonium in grassland soils and effects of protozoan grazing. PhD thesis, University of Nijmegen, The Netherlands.

Verhagen, F.J.M. and Laanbroek, H.J. (1991) Competition for ammonium between nitrifying and heterotrophic bacteria in dual-energy limited chemostats. *Applied and Environmental Microbiology* 57, 3255–3263.

Weber, A. Ph., Zwillenberg, L.O. and Van der Laan, P.A. (1952) A predacious amoeboid organism destroying larvae of the potato eelworm and other nematodes. *Nature* 169, 834–835.

Whipps, J.M. (1990) Carbon economy. In: Lynch, J.M. (ed.) *The Rhizosphere.* Wiley, Chichester, pp. 59–97.

Whipps, J.M. and Lynch, J.M. (1983) Substrate flow and utilization in the rhizosphere of cereals. *New Phytologist* 95, 605–623.

Wood, M. (1987) Predicted microbial biomass in the rhizosphere of barley in the field. *Plant and Soil* 97, 303–314.

Zak, D.R., Groffman, P.M., Pregitzer, K.S., Christensen, S. and Tiedje, J.M. (1990) The vernal dam: plant–microbe competition for nitrogen in northern hardwood forests. *Ecology* 71, 651–656.

Zwart, K.B. and Brussaard, L. (1991) Soil fauna and cereal crops. In: Firbank, L.G., Carter, N., Darbyshire, J.F. and Potts, G.R. (eds), *The Ecology of Temperate Cereal Fields.* Blackwell Scientific Publications, Oxford, pp. 139–168.

Zwart, K.B. and Darbyshire, J.F. (1992) Growth and nitrogenous excretion of a common soil flagallate, *Spumella* sp. – a laboratory experiment. *Journal of Soil Science* 43, 145–157.

Zwillenberg, L.O. (1952) *Theratromyxa weberi,* a new proteomyxan organism from soil. *Anthonie van Leeuwenhoek* 19, 101–116.

# Protozoan Interactions with the Soil Microflora and Possibilities for Biocontrol of Plant Pathogens

6

M. Pussard, C. Alabouvette and P. Levrat

*INRA Laboratoire de recherches sur la Flore pathogène et la Faune du sol, 17 rue Sully, BV 1540, 21034 Dijon Cedex, France.*

## Introduction

In soil, as in aquatic environments, protozoa are important predators of the microflora. In relation to agriculture, apart from their role in nutrient cycling discussed in Chapter 4, they may act either as beneficial biological control agents of plant diseases or as detrimental competitors to soil inocula of useful microorganisms. A much greater understanding of the interactions between soil protozoa and microflora is required before the beneficial potential of protozoa can be realized in agroecosystems.

This review is concerned not only with soil studies, but also with microbiological research in hydrobiology, where the methodological problems are not so intractable and progress has been more rapid than in soil ecology.

## Effects of Microflora on Protozoa

The soil microflora represents the main food resource of soil protozoa, but these microorganisms also produce extracellular metabolites that affect protozoa.

*Microflora as food for protozoa*

It is generally accepted that the trophic relationships between protozoa and bacteria *in situ* result in inverse relationships between population densities of bacteria and protozoa (Cutler and Crump, 1920; Cutler *et al.*, 1922; Clarholm, 1981, 1989). The diet of individual protozoan species can be wide. For example, *Arachnula impatiens* feeds by phagocytosis on bacteria, yeasts, cyanophyceae and flagellates. It is also able to perforate fungal walls, frustules of diatoms and even the integuments of nematodes to ingest the protoplasm (Old and Darbyshire, 1978). Nevertheless, the feeding of *Acanthamoeba* spp. is probably more representative of the behaviour of protozoa in general; these amoebae feed on bacteria, actinomycetes, fungal spores (Heal, 1963; Heal and Felton, 1970) and cyanophyceae (Wright *et al.*, 1978, 1981).

Depending on the nature of the microbial prey consumed, it is possible to distinguish between bacteriophagous and mycophagous protozoa. There is some prey selection, however, within these broad divisions and not all soil bacteria or fungi are devoured by any specific bacteriophagous or mycophagous protozoan, respectively (Severtzova, 1928; Singh, 1942, 1948; Dive, 1973a; Old and Patrick, 1979; Pussard *et al.*, 1980; Petz *et al.*, 1985). On the other hand, many different protozoan species have very similar diets (Singh, 1942, 1948; Anscombe and Singh, 1948; Coûteaux and Pussard, 1983). In general, Gram-negative bacteria are more edible to protozoa than Gram-positive bacteria, mycobacteria and actinomycetes. Protozoa also prefer zymogenous bacteria (*Pseudomonas, Enterobacteriaceae*) rather than autochthonous bacteria (*Arthrobacter, Micrococcus, Corynebacterium, Bacillus*), although there are several exceptions to these generalizations.

Clearly, not all the factors controlling food selection by protozoa are yet known. Some prey possess morphological features, such as extracellular capsules, increased size or projections that discourage predation (Klinge, 1958; Wilkinson, 1958; Güde, 1979; Shikano *et al.*, 1990). Some protozoa, on the other hand, have developed specialized forms of feeding that enable them to eat very large prey, such as filaments of *Oscillaria* (Comandon and de Fonbrune, 1936b) or hyphae of fungi (Comandon and de Fonbrune, 1936a; Old, 1977; Pussard *et al.*, 1979, 1980; Anderson and Patrick 1980; Petz *et al.*, 1985; Old and Chakraborty, 1986). The inedibility of some types of fungal propagule is not related to any fraction of the fungal wall, such as melanin (Bridoux, 1982).

Most of the information about feeding preferences of soil protozoa is derived from laboratory strains of protozoa, which have been isolated and cultured with a restricted range of bacteria, such as *Escherichia coli* or *Klebsiella aerogenes*. It is possible that the food preferences of soil-grown protozoa may differ greatly from those observed in mixed laboratory cultures of microflora and protozoa. Casida (1989) examined this possibility in laboratory experiments. He inoculated fresh soil samples with different bacterial species and after incubation estimated the populations of soil protozoa by the most probable number (MPN) method. Casida used the same range of bacteria as protozoan food in the MPN estimates. He found, for example, that soil samples inoculated with *Escherichia coli* had larger protozoan populations than samples inoculated with *Bacillus mycoides*. The protozoan species present were not listed, but the results suggest that some food preferences exist amongst protozoa in soil and further studies are required on this topic. Soil amendment with non-indigenous bacteria, such as *E. coli*, does not appear to encourage soil protozoa to eat indigenous soil bacteria, e.g. *Arthrobacter globiformis* or *B. mycoides*, and is probably not a method of controlling other bacteria in soil.

In general, the prey is digested within food vacuoles of protozoa. Some bacterial cells and particularly bacterial or fungal spores are able to resist digestion and are ejected alive (Oehler, 1924; Severtzova, 1928; Heal, 1963; Krishna Prasad and Gupta, 1978; Zwart and Darbyshire, 1991). In some cases, the bacteria not only survive, but grow in the food vacuoles. This has been established for *Legionella*, a bacterium responsible for legionnaires disease of humans (Skinner *et al.*, 1983; Fields *et al.*, 1984; Harf and Monteil, 1988). Recent studies show that this phenomenon is widespread and mutually advantageous. By this arrangement bacteria may avoid harsh environmental conditions (King *et al.*, 1988), and protozoa may utilize bacterial metabolites to protect themselves from mercury pollution (Mirgain *et al.*, unpublished). Parasitic unidentified bacteria have also been observed in amoebae (Drozanski, 1956, 1963).

Protozoa react to starvation by initiating sexual reproduction, morphological modifications such as transformation from microstome to macrostome in the case of some ciliates (Fenchel, 1987), encystment (Corliss and Esser, 1974), or lysis.

Some edible bacteria support less growth of protozoa than others depending on the digestibility and nutritional value of their protoplasm (Cutler and Crump, 1927; Kidder and Stuart, 1939; Singh, 1941; Burbanck, 1942; Dudziak, 1955; Curds and Vandyke, 1966; Heal and Felton, 1970; Barna and Weis, 1973; Dive, 1973a). Singh (1942) showed that strains of *K. aerogenes* from different origins, but with the same physiological characteristics, did not support the same growth rate of amoebae. The minimal prey density for protozoan growth can be as high as $10^8$ g$^{-1}$soil for some bacteria.

## Effects of microbial metabolites on protozoa

### Stimulation of the protozoan activity (excystment and growth)

Extracellular microbial metabolites play an important role in inducing protozoan excystment. Cysts carefully washed and introduced into saline solutions do not germinate (Drozanski, 1961), but excystment can be abundant in the presence of bacteria (Singh, 1941; Crump, 1950; Drozanski, 1961). Excystment of *Didinium* is induced by the presence of bacteria, but not by its prey, Paramecia (Beers, 1946; Butzel and Horwitz, 1965). Some factors, such as microbial contamination (Singh *et al.*, 1958), temperature (Dudziak, 1955), age of the cyst (Chambers and Thompson, 1974) and auto-inhibitory substances (Dubes and Jensen, 1964), probably confound the results of many such experiments. It appears that amino acids, singly or in a mixture, can induce excystment (Singh *et al.*, 1958; Drozanski, 1961; Jeffries, 1962; Kaushal and Shukla, 1977a,b). According to Kaushal and Shukla (1977c), the most active substance for cysts of *Acanthamoeba culbertsoni* is γ-aminobutyric acid, which is closely related to another very active molecule, glutamic acid (Singh *et al.*, 1958). The former is a neuro-transmitter for vertebrates and a chemical attractant of *Lithothamnium* algae for planktonic larvae of the gastropod *Haliotis rufescens* (Morse *et al.*, 1979). This example illustrates the diverse and multiple effects of some chemicals. Spore germination of the cellular slime mould, *Dictyostelium discoideum*, is induced by several factors, especially by amino acids, although under natural conditions the germination can be induced by more complex molecules from bacteria (Hashimoto *et al.*, 1976; Ihara *et al.*, 1990).

Microbial metabolites can stimulate the growth rate of protozoa, e.g. the addition of small quantities of culture filtrates of *Actinomyces griseus* and *A. aurigineus* increases the multiplication rate of Paramecia by increasing the speed at which food vacuoles are formed (Jakimova *et al.*, 1969). Similarly, culture filtrates of the fungi *Fusarium* sp. and *Gliocladium roseum* increase the rhythm of cellular division of Paramecia (Jakimova *et al.*, 1978).

*Antagonism*

Protozoan growth is inhibited on media with large nitrogen contents, even in the presence of edible bacteria (Oehler, 1924; Kurihara, 1978). Conversely, protozoa grow well on a medium with a small nitrogen content as used for *Azotobacter* and oligonitrophilic bacteria (Nikoljuk, 1965). Although, most bacteria produce toxic metabolites on nitrogen-rich media, some species or strains of microorganisms excrete particular antagonistic factors, such as pigments, toxins and antibiotics. Pigments produced by *Serratia marcescens*, *Chromobacterium violaceum* and *Pseudomonas aeruginosa* are toxic to protozoa (Singh, 1945; Groscop and Brent, 1964). According to Dive (1973b), the phenazins produced by *P. aeruginosa* and violacein produced by *Chromobacterium lividum* are the only pigments really toxic to proto-zoa. Gräf (1958) observed that 64% of *P. fluorescens* strains isolated from the natural environment produced toxic substances (peptides and organic acids), but pyoverdin pigment produced by the same bacteria was ineffective on the ciliate *Colpidium campylum* (Dive, 1973b).

Filtrates of some actinomycete cultures induce lysis of protozoa (Gel'tser, 1969; Jakimova *et al.*, 1969). Fungal strains can also have antagonistic effects on protozoa (Ebringer *et al.*, 1964; Heal and Felton, 1970; Krizkova *et al.*, 1979; Coûteaux and Dévaux, 1983). Antibiotics inhibited the phagocytic activity of the ciliate *Tetrahymena pyriformis* (Rebandel and Karpinska, 1981; Krawczynska, 1990). These examples of antagonism justify the use of the technique proposed by Singh (1946a, 1946b) to isolate protozoa from soil. This technique involves a non-nutritive medium to reduce the growth of antagonistic microorganisms and edible non-toxic bacterial prey, such as *Klebsiella aerogenes*

Dreschler (1936, 1937) described another type of antagonism, namely fungal predation of protozoa. Some fungi are able to feed on naked amoebae (*Thecamoeba terricola*) or on testate amoebae (*Euglypha* sp., *Geococcus* sp.). Infection occurs when a germ tube arising either from an adhering or ingested conidium or from a lateral branch of a vegetative hypha penetrates the protozoan protoplasm. The consequences of this fungal predation on population dynamics of protozoa have not been studied.

## Effects of Protozoa on Microflora

By feeding on the soil microflora, protozoa modify the dynamics, diversity and activity of these microbial communities.

### Population density of prey in presence of predators

When studying the survival of large bacterial inocula in soil, Alexander and coworkers observed a clear decrease in bacterial density only in unsterilized soil; they attributed this decrease to soil protozoa (Habte and Alexander, 1975; Acea and Alexander, 1988). The bacterial populations did not disappear completely; in the presence of trophic protozoa rather they stabilized at a relatively high level ($10^4$–$10^7 g^{-1}$soil), the actual level depending on the bacterial species involved (Fig. 6.1) (Danso and Alexander, 1975; Habte and Alexander, 1975, 1978a; Steinberg *et al.*, 1987). As this coexistence between protozoa and bacteria also occurred in liquid environments (Hamilton and Preslan, 1970; Seto and Tazaki, 1971; Danso *et al.*, 1975; Berk *et al.*, 1976; Sambanis and Fredrickson, 1988) as well as in soils, this equilibrium is not likely to be due entirely to some protective physical characteristic of the environment. This bacterial equilibrium has been attributed to two different phenomena: firstly, the intensity of predation would decrease when the bacterial density became low, because protozoa would catch a smaller quota of prey with the same

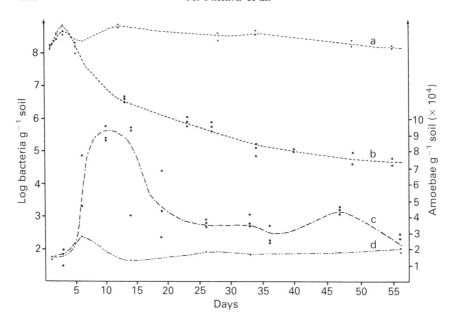

**Fig. 6.1.** Population dynamics of *Bradyrhizobium japonicum*: (a) inoculated in sterilized soil; (b) in non-sterile soil at 10⁸ bacteria g⁻¹ dry soil; and autochthonous amoebae in non-sterile soil inoculated (c) with or (d) without *B. japonicum* (Steinberg *et al.*, 1987).

level of activity (Danso *et al.*, 1975); secondly, growth of the residual bacterial population (Habte and Alexander, 1978b). The final level of the bacterial population at equilibrium can be modified by carbon supplements (inositol, mannitol) that can be assimilated by the bacteria for growth (Ramirez and Alexander, 1980). Similar results can be obtained by the addition of chemicals such as cycloheximide and thiram, that inhibit the growth of indigenous protozoa (Ramirez and Alexander, 1980; Chao and Alexander, 1981). Also, bacteria with a slow growth rate will disappear more rapidly in the presence than in the absence of a bacterial population able to support growth of indigenous protozoa (Chang and Huang, 1981; Mallory *et al.*, 1983). For example, the establishment of a population of *Rhizobium phaseoli* in soil is more difficult in the presence of large numbers of bacteria that can support a large population of protozoa (Ramirez and Alexander, 1980), than in the absence of this bacterial food source.

   Population models involving the continued coexistence of protozoa and bacterial prey are more difficult to construct than models involving carnivorous protozoa (e.g. *Didinium* with *Paramecium*). Such models are somewhat unconvincing (Sambanis and Fredrickson, 1988), because the cryptic growth of bacteria on substances excreted by protozoa (Canale *et*

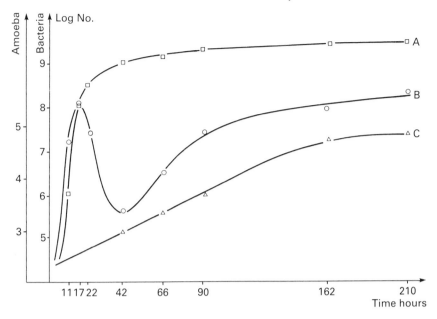

**Fig. 6.2.** Population dynamics of *Klebsiella aerogenes* introduced into sterilized soil (A) without or (B) with *Acanthomoeba castellanii* at 10⁵ bacteria g⁻¹ dry soil; (C) *A. castellanii* introduced into sterilized soil at 200 cells g⁻¹ with *K. aerogenes* (Pussard and Rouelle, 1986).

*al.*, 1973; Steinberg *et al.*, 1987) is not sufficient to explain the relatively high density of the bacterial population at equilibrium. Usually, the decline of bacterial populations in predator/prey experiments was studied after large bacterial inocula were added to soil. Under these conditions, bacteria are not able to grow profusely. It is preferable to study the growth of a bacterial population in the presence of protozoa when the two populations are introduced simultaneously into a fresh medium. Depending on the experimental conditions, the growth kinetics can be unchanged (Levrat *et al.*, 1987) or greatly modified (Pussard and Rouelle, 1986; Levrat *et al.*, 1989). The kinetics depend on the relative rates of bacterial growth and protozoan consumption of bacteria. Figure 6.2 shows a characteristic curve of the population density of bacteria in the presence of protozoa. Firstly, the bacterial population density increases, as in the control without protozoa, indicating that the effect of predation is initially extremely small. Then, as the protozoan populations increase and the bacterial population density decreases, protozoan predation becomes more evident. Finally, as a result of protozoan encystment the bacterial population increases again, because the predation is reduced. It appears that in presence of protozoa, bacterial production is greater than in the control due to the use by the bacteria of compounds released by protozoa. Under the usual nutrient-stressed conditions in soils, protozoa help to regulate the size and diversity of the bacterial communities.

Few studies have considered the relationships between protozoa and fungi. It is well established that giant amoebae belonging to the family Vampyrellidae are responsible for the perforation observed in the pigmented walls of spores of various fungi buried in soil (Old, 1977; Anderson and Patrick, 1978; Old and Darbyshire, 1978). These perforations lead to the destruction of the conidia, but the consequences of protozoan predation on the population dynamics of fungi are not clear. When the mycophagous amoeba, *Dermamoeba granifera*, was introduced into sterilized soil at the same time as a strain of *Fusarium oxysporum*, the population density of amoebae increased from $10^3$ to $25 \times 10^3$ trophozoites $g^{-1}$ soil; during the same period, the population density of the fungus was only slightly reduced below that in the control without amoebae (Alabouvette *et al.*, 1981). It has not been possible to demonstrate a decrease of the population density of *F. oxysporum* in unsterilized soil, even after stimulation of the protozoan population by addition of bacterial food (Levrat *et al.*, 1991). Couteaudier and Alabouvette (1990) also found little change in the population of *Fusarium oxysporum* after inoculation into unsterilized soil. These results suggest that, unlike bacteriophagous protozoa, the mycophagous counterparts have no significant influence on the population dynamics of fungi in natural soils.

## Effects of predation on the heterogeneity and polymorphism within soil microflora

The effects of predation on the structure of the bacterial populations and on bacterial polymorphism are still uncertain. The hypothesis proposed by Cutler and Bal (1926) that the proportion of young bacteria is greater in the presence of protozoa was not confirmed by Darbyshire (1972) in the case of *Colpoda* with *Azotobacter*. Gonzalez *et al.* (1990), however, reported that planktonic ciliates selectively consumed the biggest and most active bacteria. As a result of this predation, there was a predominance of small bacteria in the natural populations. Danso and Alexander (1975) demonstrated that bacteria surviving in the presence of protozoa are not resistant to predation, because they are consumed as soon as their population density is experimentally increased above the equilibrium level. Güde (1979) indicated that in the absence of protozoa, bacteria lose their filamentous shape or the fluffy aspect of their colonies in waste water; both features could protect these bacteria from predation by protozoa. Shikano *et al.* (1990) observed that a homogeneous bacterial population with rods of 1.5 μm in pure culture became heterogeneous in mixed culture with the ciliate

*Cyclidium.* Giant bacteria measuring up to 20 μm appeared amongst the usual short rods and these big bacteria could not be ingested by the ciliate. The giant bacteria disappeared after successive transfers in pure culture and did not reappear even if the bacteria were grown in the presence of mixed culture filtrates. This suggested that the appearance of the big bacteria was not due to chemical factors, as had been established for predator/prey relationships involving eucaryotic prey. Further observations of these phenomena are needed.

## Stimulation of bacterial metabolism by predation

As the population density of bacteria decreases in the presence of protozoa, the activity of these bacterial populations might be expected to decrease. Indeed, Russell and Hutchinson (1909) attributed the decrease of fertility in some soils to the proliferation of protozoa. Later studies have not substantiated this hypothesis and on the contrary have found a stimulation of microbial activity.

### Evidence for the stimulation of microbial metabolism

Studies reporting an increase of $CO_2$ release (Cutler and Crump, 1929; Meiklejohn, 1932; Telegdy-Kovats, 1932; Coleman *et al.*, 1978; Gupta and Germida, 1989; Henkinet *et al.*, 1990; Kuikman *et al.*, 1990) or an increased production of ammonium (Meiklejohn, 1930, 1932; Sinclair *et al.*, 1981; Woods *et al.*, 1982; Kuikman *et al.*, 1990) in mixed protozoa/microflora cultures could be attributed to protozoan metabolism as well as to increased microfloral activity. This interpretation cannot be applied when increased nitrogen fixation by *Azotobacter* was detected in mixed culture (Nasir, 1923; Cutler and Bal, 1926; Hirai and Hino, 1926, 1928; Fedorowa-Winogradowa, 1928; Hervey and Greaves, 1941; Marszewska-Ziemiecka and Malizewska, 1963; Nikolijuk, 1965, 1969a; Mavljanova, 1966; Darbyshire, 1972). The amount of nitrogen fixed was two to three times greater than in control pure cultures of *Azotobacter* (Hervey and Greaves, 1941; Darbyshire, 1972). This stimulation of bacterial activity has been confirmed in soil as well as in liquid culture for more complex interactions, such as the mineralization of organic matter (Butterfield *et al.*, 1931; Javornicki and Prokesova, 1963; Johannes, 1965; Curds *et al.*, 1968; Stout, 1973; Barsdate *et al.*, 1974; Fenchel and Harrison, 1976; Fenchel, 1977, 1987; Anderson *et al.*, 1978, 1981; Cole *et al.*, 1978; Coleman *et al.*, 1978; Woods *et al.*, 1982; Güde, 1985; Griffiths, 1986, 1989).

*Mechanisms of stimulation*

Four hypotheses have been proposed to explain the stimulation of bacterial activity by protozoa.

1.  Protozoan predation maintained a higher proportion of young and active bacteria in the population (Cutler and Bal, 1926), either by consuming preferentially the old bacteria or by delaying the decline of the metabolism that appears when the population density reaches the carrying capacity of the medium (Butterfield and Purdy, 1931; Johannes, 1965; Curds, 1977).

2.  Protozoa make the medium more favourable for bacterial activity either by consuming oxygen and so preventing high concentrations of oxygen that inhibit nitrogen fixation by *Azotobacter* (Darbyshire, 1972) or by excreting $NH_4$ and so preventing the acidification of the environment (Hirai and Hino, 1926).

3.  Protozoa produce substances that stimulate the growth and activity of bacteria (Hervey and Greaves, 1941; Straskrabova-Prokesova and Legner, 1966; Nikoljuk, 1969a, 1969b). This phenomenon is promoted by γ-irradiation as a result of increased permeability of protozoan membranes (Nikoljuk, 1965; Mavljanova, 1966). Darbyshire (1972) rejected this hypothesis, because it does not explain the effect of temperature on the stimulation of nitrogen fixation in mixed culture of *Colpoda* and *Azotobacter*.

4.  The explanation usually favoured today is that concerned with the recycling of the biomass consumed by protozoa. Amoebae assimilate only 58% of the consumed biomass and 42% is rejected as excretory products (Heal, 1967). As protozoa feed on bacteria having C:N and C:P ratios close to their own contents they need to eliminate excess nitrogen and phosphorus to compensate for the losses in carbon due to respiration. Nitrogen is excreted in the form of $NH_4$ (Hunt *et al.*, 1977; Woods *et al.*, 1982; Clarholm, 1985) or organic compounds (Palka, 1988); phosphorus is also excreted as mineral or organic compounds (Johannes, 1965; Taylor, 1986). Therefore, in mixed culture with protozoa, more nutrients should be available for bacteria than in pure bacterial cultures (Stout, 1973; Pussard and Rouelle, 1986). Nevertheless, this increased growth of bacteria would not be detectable, because at the same time protozoa would be continually eating the bacteria. Coûteaux *et al.* (1988) termed this growth as cryptic bacterial growth. Even if the results of Sambanis *et al.* (1987) seem to confirm this hypothesis, the recycling of biomass cannot explain the high level of stimulation observed in some cases (Pussard and Rouelle, 1986; Kuikman, 1990), but many scientists have not considered the other hypotheses.

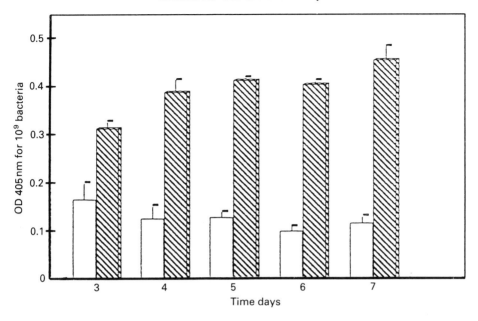

**Fig. 6.3.** Pyoverdin concentration (optical density at 405 nm) calculated for $10^9$ *Pseudomonas* ml⁻¹ in King's B medium diluted ten times amended with 10% of the filtrate of an axenic culture of amoebae (hatched columns); amended with 10% of pure King's B medium as control (white columns) (Levrat *et al.*, 1992).

*New evidence of protozoan byproducts stimulating microbial activity*

In *Pseudomonas fluorescens/Acanthamoeba castellanii* mixed cultures, the enhancement of bacterial activity was assessed by the production of specific bacterial metabolites, such as the fluorescent pigment, pyoverdin, and cyanidric acid. Cultured on liquid King's B medium diluted ten times to allow for simultaneous growth of bacteria and amoebae, *Pseudomonas* produced twice as much pigment in mixed culture as in pure culture (Levrat *et al.*, 1989). The same stimulation was observed after amendment of a pure culture of *Pseudomonas* with 2–5% of culture filtrates from an 8-day-old mixed culture. Under these experimental conditions, the enhancement of bacterial activity is not related to any increase of population density of the bacteria (Levrat, 1990; Levrat *et al.*, 1992). The same type of result was obtained with another amoeba, *Naegleria gruberi*, which is not closely related to *Acanthamoeba*. The stimulation of bacterial activity was also observed after addition of culture filtrates from axenic *A. castellanii* grown on King's B medium to the bacterial cultures (Fig. 6.3). The experimental procedures used suggest that this increased production of pigments was not connected

*M. Pussard* et al.

with iron deficiency. The amount of cyanidric acid produced by *Pseudo-monas* was also 2 to 3 times greater after addition of culture filtrate obtained from axenic culture of *A. castellanii* (Levrat, unpublished). Under the same experimental conditions, bacterial respiration increased by 20–25% and $NH_4$ production was stimulated. In the absence of protozoa in the culture filtrate, the increased production of $CO_2$ and $NH_4$ was due to an enhancement of bacterial metabolism (Levrat, 1990). These results suggested that amoebae produced one or more substances that stimulated bacterial metabolism. It was concluded that bacteriophagous amoebae not only recycle the bacterial biomass, but can produce specific chemicals that stimulate the bacterial activity even at low concentrations. This stimulation is not due to a modification of the growth rate of the population, as stated by Hervey and Greaves (1941) and Nikoljuk (1965). Therefore, it appears that the substances produced by protozoa modify bacterial metabolism. The increased production of ammonium, phosphates and sulfates in the presence of protozoa does not result from greater excretion by protozoa, but from an increased activity of the bacteria. This interpretation is in agreement with the results of Barsdate *et al.* (1974), but disagrees with the opinion of scientists who subscribed to Johannes' (1965) suggestion that it results from biomass recycling (Clarholm, 1981, 1985; Woods *et al.*, 1982; Coûteaux *et al.*, 1988; Gupta and Germida, 1989). If this form of stimulation is common amongst protozoa, it would increase the importance of protozoa in many environments and the association of a protozoan predator with bacterial prey would be more important than the separate roles of each participant (cf. concept of the module of Thomas, 1990). Protozoa would then seem to behave as 'prudent predators' (Slobodkin, 1968), because by stimulating the metabolic activity of the bacterial population on which they feed, they would increase the chances of survival of their prey. Predation would become mutually advantageous to predator and prey.

## Application to Biological Control

By analogy with other invertebrate predators and pests, it is possible that protozoa could limit crop damage caused by phytopathogenic microorganisms in soils and hydroponic systems. This possibility has been reviewed by Foissner (1987). Two different aspects of biocontrol are considered in the following discussion; that associated with the consumption of microbial prey and that associated with the stimulation of bacterial metabolism.

## Direct effect of predation and consumption of pathogens

It seems reasonable to suppose that by grazing on pathogenic microorganisms, protozoa could limit the incidence of disease. Hino (1935) proposed the use of protozoa to control soil-borne plant pathogens. Esser *et al.* (1975) also suggested the use of mycophagous amoebae to control fusarium wilts. Old and Patrick (1979) suggested a similar role for mycophagous giant amoebae. These latter amoebae are the only organisms known to perforate the pigmented walls of conidia and hyphae of soil fungi. Chakraborty and Warcup (1983, 1984) observed that mycophagous amoebae were more abundant in suppressive rather than conducive soils with regard to *Gaeumannomyces graminis* and they suggested that these giant amoebae were involved in this disease suppression. Homma *et al.* (1979) and Chakraborty (1985) observed that pigmented hyphae disappeared more quickly in suppressive than in conducive soils, whilst the hyaline hyphae disappeared at the same speed in both soils. The amoebae obviously play some part in the destruction of this pathogen in soil, but many other microbial interactions have been proposed as mechanisms of take-all decline (Cook and Baker, 1983). As the higher density of amoebae is correlated with a higher density of the microflora including the antagonistic bacteria, the suppressiveness of the soil could also be related to greater bacterial antagonism. The results published by Chakraborty *et al.* (1985) can be interpreted in this manner. They also showed a reduction of root colonization of *Pinus radiata* by the mycorrhizal fungus *Rhizopogon luteolus* in the presence of mycophagous amoebae. This reduction can be alternatively attributed to the stimulation of antagonism by *Klebsiella* spp. on the fungus, as these bacteria were the prey of the amoebae.

As discussed earlier, in sterilized soil *Dermamoeba granifera* produced only a slight decrease of the population density of *Fusarium oxysporum* in spite of being able to feed on this fungus (Alabouvette *et al.*, 1981). In unsterilized soil, the inoculum density of *Fusarium oxysporum* remained stable for at least one month after its introduction at the initial concentration of $10^6$ CFU g$^{-1}$ soil (Levrat *et al.*, 1991). It is difficult to envisage a reduction in wilt disease by *Fusarium* where there is no significant reduction in the size of pathogenic inocula. Only 10–100 propagules of pathogenic *Fusarium oxysporum* per gram of soil are required to induce a severe disease in a susceptible crop. The activities of amoebae are severely restricted by environmental factors, e.g. to very wet soils, whereas fungi are still active at low water potentials. The lower limit of –20 kPa for the giant amoebae is wetter than many soils at field capacity (Cook and Homma, 1979). This represents a serious limitation to the usefulness of amoebae as antagonists for fungal diseases (Cook and Baker, 1983).

To our knowledge, since the early work of Habte and Alexander (1975), nothing has been published concerning the relationships between protozoa and plant pathogenic bacteria. These predator/prey relationships result in dynamic equilibria; the bacterial populations are usually stabilized at a density larger than $10^4$ CFU $g^{-1}$ soil. This concentration is generally larger than the thresholds required to induce severe diseases. Direct consumption of pathogenic bacteria by protozoa is therefore not likely to reduce soil inocula sufficiently to limit the incidence of bacterial plant diseases.

### Effects of substances excreted by protozoa

As early as 1941, Brodsky observed that the ciliate *Colpoda* was antagonistic to *Verticillium dahliae in vitro* and suggested that this antagonism was due to the production of a lytic factor. Nikoljuk (1965) made similar observations and extended them to another plant pathogenic fungus, *Rhizoctonia solani. Pseudomonas putida* has been suggested as a biocontrol agent for fusarium wilt (Baker *et al.*, 1986). Therefore, it seemed worthwhile to study the effect of the bacteriophagous amoeba, *Acanthamoeba castellanii*, on the antagonism exerted by *Pseudomonas putida* on *Fusarium oxysporum* in heat-treated soil (Levrat *et al.*, 1987). In the presence of amoebae, the population density of *F. oxysporum* is significantly lower than in the control without amoebae. It was verified that this decrease of the inoculum density was not due to the direct consumption of fungal spores by the amoebae. In culture filtrates of mixed culture (*P. putida* and *A. castellanii*), the percentage germination of fungal conidia was significantly reduced and the growth rates of the germ-tubes were also reduced in comparison with the control (culture filtrate of *P. putida* alone) (Levrat, unpublished). These results showed that protozoa increased the antagonism of *P. putida* on *F. oxysporum*. They also open the way to another form of biological control: the use of culture filtrates of protozoa or the direct utilization of specific chemicals produced by protozoa to stimulate the antagonistic activity of bacteria.

## Conclusions

This review of the literature shows that many studies have been concerned with the population dynamics of bacteria in the presence of protozoa, whereas only a few scientists have investigated chemical interactions between microflora and protozoa. In interactions between *P. fluorescens* and *A. castellanii*, it has been demonstrated that bacteriophagous amoebae produce specific chemicals that stimulate the metabolism of the bacterial population (Levrat, 1990). The significance of this phenomenon has not been assessed fully, but we think that it could be of general importance (Pussard, 1991). This review also shows that the study of the interactions between protozoa and fungi has been neglected. Protozoan feeding on the microflora in soil did not result in the complete suppression of the prey, so it seems unlikely that protozoa can be used directly to control the numbers of plant pathogenic bacteria or fungi. The demonstration that protozoa can stimulate bacterial antagonism, however, does provide a novel approach to the biological control of plant diseases.

## Acknowledgements

The authors wish to thank Dr J.F. Darbyshire for his advice on the manuscript, and M. Janisz and P. Richard for help with references and secretarial assistance.

## References

Acea, M.J. and Alexander, M. (1988) Growth and survival of bacteria introduced into carbon-amended soil. *Soil Biology and Biochemistry* 20, 703–709.

Alabouvette, C., Lemaitre, I. and Pussard, M. (1981) Densité de population de l'amibe mycophage *Thecamoeba granifera* s.sp. *minor* (Amoebida, Protozoa). Mesure et variations expérimentales dans le sol. *Revue d'Écologie et de Biologie du Sol* 18, 179–192.

Anderson, R.V., Elliott, E.T., McClellan, J.F., Coleman, D.C., Cole, C.V. and Hunt, H.W. (1978) Trophic interactions in soils as they affect energy and nutrient dynamics. III – Biotic interactions of bacteria, amoebae and nematodes. *Microbial Ecology* 4, 361–371.

Anderson, R.V., Coleman, D.C. and Cole, C.V. (1981) Effects of saprotrophic grazing on net mineralization. *Ecological Bulletin* 33, 201–216.

Anderson, T.R. and Patrick, Z.A. (1978) Mycophagous amoeboid organisms from soil that perforate spores of *Thielaviopsis basicola* and *Cochliobolus sativus*. *Phytopathology* 68, 1618–1626.

Anderson, T.R. and Patrick, Z.A. (1980) Soil vampyrellid amoebae that cause small perforations in conidia of *Cochliobolus sativus*. *Soil Biology and Biochemistry* 12, 159–167.

Anscombe, F.J. and Singh, B.N. (1948) Limitation of bacteria by micropredation in soil (Amoebae). *Nature* 161, 140.

Baker, R., Elad, Y. and Sneh, B. (1986) Physical, biological and host factors in iron competition in soils. In: Swinburne, T.R. (ed.), *Iron, Siderophores and Plant Diseases*. Plenum Publishing Co., New York, pp. 77–84.

Barna, I. and Weis, D.S. (1973) The utilization of bacteria as food for *Paramecium bursaria*. *Transactions of the American Microscopical Society* 92, 434–440.

Barsdate, R.J., Prentki, R.T. and Fenchel, T. (1974) Phosphorus cycle of model ecosystem: significance for decomposer food chains and effect of bacterial grazers. *Oikos* 25, 239–251.

Beers, C.D. (1946) Excystment in *Didinium nasutum* with special reference to the role of bacteria. *Journal of Experimental Zoology* 103, 201–231.

Berk, S.G., Colwell, R.R. and Small, E.B. (1976) A study of feeding responses to bacterial prey by estuarine ciliates. *Transactions of the American Microscopical Society* 95, 514–520.

Bridoux, P. (1982) Etude de deux amibes mycophages. Relations trophiques et altérations ultra-structurales des organes fongiques. DEA de Biologie et Physiologie végétales, Université de Nancy I, UER, PCB.

Brodsky, A.L. (1941) Antagonistische Beziehungen zwischen den Bodeninfusorien und dem pathogenen Bodenpilz. *Comptes Rendus de l'Académie des Sciences de l'URSS, Moscow* 33, 81–83.

Burbanck, W.D. (1942) Physiology of the ciliate *Colpidium colpoda*. I – Effect of various bacteria as food on the division rate of *Colpidium colpoda*. *Physiological Zoology* 15, 342–362.

Butterfield, C.T. and Purdy, W.C. (1931) Some interrelationships of plankton and bacteria in natural purification of polluted water. *Industrial and Engineering Chemistry* 23, 213–218.

Butterfield, C.T., Purdy, W.C. and Theriault, E.J. (1931) Studies on natural purification in polluted waters. IV – The influence of plankton on the biochemical oxidation of organic matter. *Public Health Reports* 46, 393–426.

Butzel, H.M. and Horwitz, H. (1965) Excystment of *Didinium nasutum*. *Journal of Protozoology* 12, 413–416.

Canale, R.P., Lustig, T.D., Kehrberger, P.M. and Salo, J.E. (1973) Experimental and mathematical modeling studies of protozoan predation on bacteria. *Biotechnology and Bioengineering* 15, 707–728.

Casida, L.E. (1989) Protozoan response to the addition of bacterial predators and other bacteria to soil. *Applied and Environmental Microbiology* 55, 1857–1859.

Chakraborty, S. (1985) Survival of wheat take-all fungus in suppressive and non-suppressive soils. *Pedobiologia* 28, 13–18.

Chakraborty, S. and Warcup, J.H. (1983) Soil amoebae and saprophytic survival of *Gaeumannomyces graminis tritici* in a suppressive pasture soil. *Soil Biology and Biochemistry* 15, 181–185.

Chakraborty, S. and Warcup, J.H. (1984) Populations of mycophagous and other amoebae in take-all suppressive and non-suppressive soils. *Soil Biology and Biochemistry* 16, 197–199.

Chakraborty, S., Theodorou, C. and Bowen, G.D. (1985) The reduction of root colonization by mycorrhizal fungi by mycophagous amoebae. *Canadian Journal of Microbiology* 31, 295–297.

Chambers, J.A. and Thompson, J.E. (1974) Age dependent excystment of the protozoan *Acanthamoeba castellanii*. *Journal of General Microbiology* 80, 375–380.

Chang, M.C. and Huang, T.C. (1981) Effects of the predation of *Tetrahymena pyriformis* on the population of *Aeromonas hydrophila*. *National Science Council Monthly. ROC* 9, 552–556.

Chao, W.L. and Alexander, M. (1981) Interaction between protozoa and *Rhizobium* in chemically amended soil. *Soil Science Society of America Journal* 45, 48–50.

Clarholm, M. (1981) Protozoan grazing of bacteria in soil. Impact and importance. *Microbial Ecology* 7, 343–350.

Clarholm, M. (1985) Interactions of bacteria, protozoa and plants leading to mineralization of soil nitrogen. *Soil Biology and Biochemistry* 17, 181–187.

Clarholm, M. (1989) Effects of plant–bacterial–amoebal interactions on plant uptake of nitrogen under field conditions. *Biology and Fertility of Soils* 8, 373–378.

Cole, C.V., Elliott, E.T., Hunt, H.W. and Coleman, D.C. (1978) Trophic interactions in soils as they affect energy and nutrient dynamics. V – Phosphorus transformations. *Microbial Ecology* 4, 381–387.

Coleman, D.C., Anderson, R.V., Cole, C.V., Elliott, E.T., Woods, L. and Campion,

M.K. (1978) Trophic interactions in soils as they affect energy and nutrient dynamics. IV – Flows of metabolic and biomass carbon. *Microbial Ecology* 4, 373–380.

Comandon, J. and Fonbrune, P. de (1936a) Amibe mycophage. Enregistrement cinématographique. *Comptes Rendus des Séances de la Société de Biologie et de ses Filiales, Paris* 123, 1069.

Comandon, J. and Fonbrune, P. de (1936b) Mécanisme de l'ingestion d'oscillaires par des amibes. Enregistrement cinématographique. *Comptes Rendus des Séances de la Société de Biologie et de ses Filiales, Paris* 123, 1170.

Cook, R.J. and Baker, K.F. (1983) *The Nature and Practice of Biological Control of Plant Pathogens.* American Phytopathological Society, St Paul.

Cook, R.J. and Homma, Y. (1979) Influence of soil water potential on activity of amoebae responsible for perforations of fungal spores. *Phytopathology* 69, 914.

Corliss, J.O. and Esser, S.C. (1974) Comments on the role of the cyst in the life cycle and survival of free-living protozoa. *Transactions of the American Microscopical Society* 93, 578–593.

Couteaudier, Y. and Alabouvette, C. (1990) Survival and inoculum potential of conidia and chlamydospores of *Fusarium oxysporum* f.sp. *lini* in soil. *Canadian Journal of Microbiology* 36, 551–556.

Coûteaux, M.M. and Dévaux, J. (1983) Effet d'un enrichissement en champignons sur la dynamique d'un peuplement thécamoebien d'un humus. *Revue d'Écologie et de Biologie du Sol* 20, 519–545.

Coûteaux, M.M. and Pussard, M. (1983) Nature du régime alimentaire des protoaires du sol. In: Lebrun, P., André, H.M., De Medts, A., Grégoire-Wibo, C. and Wauthy, G. (eds), *New Trends in Soil Biology.* Dieu-Brichart-Ottignies-Louvain la Neuve, Belgique, pp. 179–195.

Coûteaux, M.M., Faurie, G., Palka, L. and Steinberg, C. (1988) La relation prédateur–proie (protozoaires–bactéries) dans les sols: rôle dans la régulation des populations et conséquences sur les cycles du carbone et de l'azote. *Revue d'Écologie et de Biologie du Sol* 25, 1–31.

Crump, L. (1950) The influence of the bacterial environment on the excystment of amoebae from soil. *Journal of General Microbiology* 4, 16–21.

Curds, C.R. (1977) Microbial interactions involving protozoa. *Society for Applied Bacteriology, Symposium Series* 6, 69–105.

Curds, C.R. and Vandyke, J.M. (1966) The feeding habits and growth rates of some freshwater ciliates found in activated-sludge plants. *Journal of Applied Ecology* 3, 127–137.

Curds, C.R., Cockburn, A. and Vandyke, J.M. (1968) An experimental study of the role of protozoa in the activated sludge process. *Water Pollution Control* 8, 312–329.

Cutler, D.W. and Bal, D.V. (1926) Influence of protozoa on the process of nitrogen fixation by *Azotobacter chroococcum. Annals of Applied Biology* 13, 516–534.

Cutler, D.W. and Crump, L.M. (1920) Daily periodicity in the numbers of active soil flagellates: with a brief note on the relation of trophic amoebae and bacterial numbers. *Annals of Applied Biology* 7, 11–24.

Cutler, D.W. and Crump, L.M. (1927) The qualitative and quantitative effects of food on the growth of a soil amoeba (*Hartmannella hyalina*). *British Journal of Experimental Biology* 5, 155–165.

Cutler, D.W. and Crump, L.M. (1929) Carbon dioxide production in sands and soils in the presence and absence of amoebae. *Annals of Applied Biology* 16, 472–482.

Cutler, D.W., Crump, L.M. and Sandon, H. (1922) A quantitative investigation of the bacterial and protozoan population of the soil, with an account of the protozoan fauna. *Philosophical Transactions of the Royal Society of London. Series B* 211, 317–350.

Danso, S.K.A. and Alexander, M. (1975) Regulation of predation by prey density: the protozoan–*Rhizobium* relationship. *Applied Microbiology* 29, 515–521.

Danso, S.K.A., Keya, S.O. and Alexander, M. (1975) Protozoa and the decline of *Rhizobium* population added to soil. *Canadian Journal of Microbiology* 21, 884–895.

Darbyshire, J.F. (1972) Nitrogen fixation by *Azotobacter chroococcum* in the presence of *Colpoda steini*. I – The influence of temperature. *Soil Biology and Biochemistry* 4, 359–369.

Dive, D. (1973a) Nutrition holozoïque de *Colpidium campylum*. Phénomènes de sélection et d'antagonisme avec les bactéries. *Water Research* 7, 695–706.

Dive, D. (1973b) La nutrition holozoïque des protozoaires ciliés. Ses conséquences dans l'épuration naturelle et artificielle. *L'Année Biologique* 12, 343–380.

Drechsler, C. (1936) A Fusarium-like species of *Dactylella* capturing and consuming testaceous rhizopods. *Journal of the Washington Academy of Sciences* 26, 397–404.

Drechsler, C. (1937) New Zoopagaceae destructive of soil rhizopods. *Mycologia* 29, 229–249.

Drozanski, W. (1956) Fatal bacterial infection in soil amoebae. *Acta Microbiologica Polonica* 5, 315–317.

Drozanski, W. (1961) The influence of bacteria on the excystment of soil amoebae. *Acta Microbiologica Polonica* 10, 147–153.

Drozanski, W. (1963) Observations on intracellular infection of amoebae by bacteria. *Acta Microbiologica Polonica* 12, 9–24.

Dubes, G.R. and Jensen, T. (1964) Production and properties of an excystment inhibitor from *Acanthamoeba* sp. *Journal of Parasitology* 50, 380–385.

Dudziak, B. (1955) The influence of the temperature and of the bacterial environment on the excystment and growth of soil amoeba. *Acta Microbiologica Polonica* 4, 115–125.

Ebringer, L., Balan, J. and Nemec, P. (1964) Incidence of antiprotozoal substances in Aspergillaceae. *Journal of Protozoology* 11, 153–156.

Esser, R.P., Ridings, W.H. and Sobers, E.K. (1975) Ingestion of fungus spores by protozoa. *Proceedings of the Soil and Crop Sciences of Florida* 34, 206–208.

Fedorowa-Winogradowa, T. (1928) Beiträge zur Frage der Wirkung der Bodenamöben auf das Wachstum und die Entwicklung des *Azotobacter chroococcum* unter Versuchsbedingungen auf sterilem Boden. *Centralblatt für Bakteriologie, II* 74, 14–22.

Fenchel, T. (1977) The significance of bactivorous protozoa in the microbial community of detrital particles. In: Cairns, J. (ed.), *Aquatic Microbial Communities*. Garland Publishers, New York, London, pp. 529–543.

Fenchel, T. (1987) *Ecology of Protozoa*. Science Tech, Madison.

Fenchel, T. and Harrison, P. (1976) The significance of bacterial grazing and mineral cycling for the decomposition of particulate detritus. In: Anderson, J.M. and MacFadyen, A. (eds), *The Role of Terrestrial and Aquatic Organisms in Decomposition Processes*. Blackwell Scientific Publications, Oxford, pp. 285–299.

Fields, B.S., Shotts, E.B., Feeley, J.C., Gorman, G.W. and Martin, W.T. (1984) Proliferation of *Legionella pneumophila* as an intracellular parasite of the ciliated protozoan *Tetrahymena pyriformis*. *Applied and Environmental Microbiology* 47, 467–471.

Foissner, W. (1987) Soil protozoa: fundamental problems, ecological significance, adaptations in ciliates and testaceans, bioindicators and guide to the literature. *Progress in Protistology* 2, 69–212.

Gel'tser, J.G. (1969) Protistocidic activity of the soil microflora. In: Strelkov A.A., Sukhanova, K.M. and Raikov, I.B. (eds), *Progress in Protozoology*. Nauka, Leningrad, p. 192.

Gonzalez, J.M., Sherr, E.B. and Sherr, B.F. (1990) Size-relative grazing on bacteria by natural assemblages of estuarine flagellates and ciliates. *Applied and Environmental Microbiology* 56, 583–589.

Gräf, W. (1958) Über Gewinnung und Beschaffenheit protostatisch-protocider Stoffe aus Kulturen von *Pseudomonas fluorescens*. *Archiv für Hygiene* 142, 267–275.

Griffiths, B.S. (1986) Mineralization of nitrogen and phosphorus by mixed cultures of the ciliate protozoan *Colpoda steinii*, the nematode *Rhabditis* sp. and the bacterium *Pseudomonas fluorescens*. *Soil Biology and Biochemistry* 18, 637–641.

Griffiths, B.S. (1989) Enhanced nitrification in the presence of bacteriophagous protozoa. *Soil Biology and Biochemistry* 21, 1045–1051.

Groscop, J.A. and Brent, M.M. (1964) The effects of selected strains of pigmented microorganisms on small free-living amoebae. *Canadian Journal of Microbiology* 10, 579–584.

Güde, H. (1979) Grazing by protozoa as selection factor for activated sludge bacteria. *Microbial Ecology* 5, 225–237.

Güde, H. (1985) Influence of phagotrophic processes on the regeneration of nutrients in two-stage continuous culture systems. *Microbial Ecology* 11, 193–204.

Gupta, V.V.S.R. and Germida, J.J. (1989) Influence of bacterial–amoebal interactions on sulfur transformations in soil. *Soil Biology and Biochemistry* 21, 921–930.

Habte, M. and Alexander, M. (1975) Protozoa as agents responsible for the decline of *Xanthomonas campestris* in soil. *Applied Microbiology* 29, 159–164.

Habte, M. and Alexander, M. (1978a) Protozoan density and the coexistence of protozoan predators and bacterial prey. *Ecology* 59, 140–146.

Habte, M. and Alexander, M. (1978b) Mechanisms of persistence of low numbers of bacteria preyed upon by protozoa. *Soil Biology and Biochemistry* 10, 1–6.

Hamilton, R. and Preslan, J. (1970) Observations on the continuous culture of a planktonic phagotrophic protozoan. *Journal of Experimental Marine Biology and Ecology* 5, 94–104.

Harf, C. and Monteil, H. (1988) Interactions between free-living amoebae and *Legionella* in the environment. *Water Science and Technology* 20, 235–239.

Hashimoto, Y., Tanaka, Y. and Yamada, T. (1976) Spore germination promoter of *Dictyostelium discoideum* excreted by *Aerobacter aerogenes*. *Journal of Cell Science* 21, 261–271.

Heal, O.W. (1963) Soil fungi as food for amoebae. In: Doeksen, J. and Van Der Drift, J. (eds), *Soil Organisms*. North-Holland Publishing Company, Amsterdam, pp. 289–297.

Heal, O.W. (1967) Quantitative feeding studies on soil amoebae. In: Graff, O. and Satchell, J.E. (eds), *Progress in Soil Biology*. Friedr. Vieweg & Sohn GmbH, Braunschweig, pp. 120–126.

Heal, O.W. and Felton, M.J. (1970) Soil amoebae: their food and their reaction to microflora exudates. In: Watson, A. (ed.), *Animal Populations in Relation to their Food Resources*. Blackwell Scientific Publications, Oxford, pp. 145–162.

Henkinet, R., Coûteaux, M.M., Billes, G., Bottner, P. and Palka, L. (1990) Acceleration du turnover du carbone et stimulation du priming effect par la predation dans un humus forestier. *Soil Biology and Biochemistry* 22, 555–561.

Hervey, R.J. and Greaves, J.E. (1941) Nitrogen fixation by *Azotobacter chroococcum* in the presence of soil protozoa. *Soil Science* 51, 85–100.

Hino, I. (1935) Antagonistic action of soil microbes with special reference to plant hygiene. *Transactions 3rd International Congress of Soil Science, Oxford*, 173–174.

Hirai, K. and Hino, I. (1926) Studies on soil protozoa. I – Influence of soil protozoa on nitrogen fixation of *Azotobacter*. *Bulletin of the Agricultural and Chemical Society of Japan* 2, 85–86.

Hirai, K. and Hino, I. (1928) Influence of soil protozoa on nitrogen fixation by *Azotobacter*. *Proceedings and Papers of the First International Congress of Soil Science, Washington* 3, 160–165.

Homma, Y., Sitton, J.W., Cook, R.J. and Old, K.M. (1979) Perforation and destruction of pigmented hyphae of *Gaeumannomyces graminis* by vampyrellid amoebae from Pacific North-West wheat field soils. *Phytopathology* 69, 1118–1122.

Hunt, J.W., Cole, C.V., Klein, D.A. and Coleman, D.C. (1977) A simulation model for the effect of predation on continuous culture. *Microbial Ecology* 3, 259–278.

Ihara, M., Tanaka, Y., Hashimoto, Y. and Yanagisawa, K. (1990) Partial purification and some properties of spore germination promoter (S) for *Dictyostelium discoideum* excreted by *Klebsiella aerogenes*. *Agricultural and Biological Chemistry, Tokyo* 54, 2619–2627.

Jakimova, G.I., Borisova, T.A. and Rasumovsky, P.N. (1969) Einfluss von Mikrobenmetaboliten auf die Nahrungsaufnahme bei *Paramecium caudatum*. In: Strelkov, A.A., Sukhanova, K.M. and Raikov, I.B. (eds), *Progress in Protozoology*. Nauka, Leningrad, pp. 144–145.

Jakimova, G.I., Borisova, T.A., Garkavenko, A.I. and Rasumovsky, P.N. (1978) The influence of fungi and actinomycetes filtrates on the rate of *Paramecium caudatum* fission. *Protozoologiya* 3, 123–133.

Javornicki, P. and Prokesova, V. (1963) The influence of protozoa and bacteria upon the oxidation of organic substances in water. *Internationale Revue der gesamten Hydrobiologie* 48, 335–350.

Jeffries, W.B. (1962) Studies on specific chemicals as excysting agents for *Pleurotricha lanceolata*. *Journal of Protozoology* 9, 375–376.

Johannes, R.E. (1965) Influence of marine protozoa on nutrient regeneration. *Limnology and Oceanography* 10, 434–442.

Kaushal, D.C. and Shukla, O.P. (1977a) Excystment of axenically prepared cysts of *Hartmannella culbertsoni*. *Journal of General Microbiology* 98, 117–123.

Kaushal, D.C. and Shukla, O.P. (1977b) Characterization of excystment factors from an aqueous extract of *Escherichia coli*. *Indian Journal of Experimental Biology* 15, 538–541.

Kaushal, D.C. and Shukla, O.P. (1977c) Structure activity relationship of excystment agents of *Hartmannella culbertsoni*. *Indian Journal of Experimental Biology* 15, 542–543.

Kidder, A.W. and Stuart, C.A. (1939) Growth studies on ciliates. I – The role of bacteria in the growth and reproduction of *Colpoda*. *Physiological Zoology* 12, 329–340.

King, C.H., Shotts, E.B. Wooley, R.E. and Porter, K.G. (1988) Survival of coliforms and bacterial pathogens within protozoa during chlorination. *Applied and Environmental Microbiology* 54, 3023–3033.

Klinge, K. (1958) *Pseudomonas fluorescens*-polysaccharide als Schutz gegen Phagozytose durch Amöben. *Naturwissenschaften* 45, 550–551.

Krawczynska, W. (1990) Polymyxin-B, Gentamycin and Neomycin inhibit phago-cytic activity of *Tetrahymena pyriformis*. *Acta Protozoologica* 29, 195.

Krishna Prasad, B.N. and Gupta, S.K. (1978) Preliminary report on the engulfment and retention of mycobacteria by trophozoites of axenically grown *Acanthamoeba castellanii* Douglas, 1930. *Current Science* 47, 245–247.

Krizkova, L., Balanova, J. and Balan, J. (1979) Incidence of antiprotozoal antivermal activities in Imperfecti Fungi collected in the People's Republic of Mongolia. *Biologia* 34, 241–245.

Kuikman, P.J. (1990) Mineralization of nitrogen by protozoan activity in soil. PhD Thesis, University of Wageningen, The Netherlands.

Kuikman, P.J., Van Elsas, J.D., Jansen, A.G., Burgers, S.L.G.E. and Van Veen, J.A. (1990) Population dynamics and activity of bacteria and protozoa in relation to their spatial distribution in soil. *Soil Biology and Biochemistry* 22, 1063–1073.

Kurihara, Y. (1978) Studies of the interaction in a microcosm. *Science Reports of Tohoku University, Series IV (Biology)* 37, 161–177.

Levrat, P. (1990) Contribution à l'étude des interactions entre protozoaires et micro-flore du sol: effet d'une amibe bactériophage *Acanthamoeba castellanii* sur le métabolisme de *Pseudomonas* fluorescents. PhD Thesis, University of Claude Bernard, Lyon.

Levrat, P., Alabouvette, C. and Pussard, M. (1987) Rôle des protozoaires dans les systèmes interactifs: cas des relations bactéries-champignons. *Revue d'Écologie et de Biologie du Sol* 24, 503–514.

Levrat, P., Pussard, M. and Alabouvette, C. (1989) Action d'*Acanthamoeba castel-lanii* (Protozoa: Amoebida) sur la production de sidérophores par la bactérie *Pseudomonas putida*. *Comptes Rendus de l'Académie des Sciences, Paris* 308, ser. III, 161–164.

Levrat, P., Pussard, M., Steinberg, C. and Alabouvette, C. (1991) Regulation of *Fusarium oxysporum* populations introduced into soil: the amoebal predation hypothesis. *FEMS Microbiology Ecology* 86, 123–130.

Levrat, P., Pussard, M. and Alabouvette, C. (1992) Enhanced bacterial metabolism of a *Pseudomonas* strain in response to the addition of culture filtrate of a bacteriophagous amoeba. *European Journal of Protistology* 28, 79–84.

Mallory, L.M., Yuk, C.S., Liang, L.N. and Alexander, M. (1983) Alternative prey: a mechanism for elimination of bacterial species by protozoa. *Applied and Environmental Microbiology* 46, 1073–1079.

Marszewska-Ziemiecka, J. and Malizewska, W. (1963) [Mutual relationship between *Azotobacter* and soil amoebae] *Pamietnik Pulawskii* 11, 35–48 (in Polish).

Mavljanova, M.I. (1966) Über die Reaktion von Infusorien der Gattung *Colpoda* auf Gamma-Strahlen von $Co^{60.}$ *Pedobiologia* 6, 129–139.

Meiklejohn, J. (1930) The relation between the numbers of a soil bacterium and the ammonia produced by it in peptone solutions; with some reference to the effect on this process of the presence of amoebae. *Annals of Applied Biology* 17, 614–637.

Meiklejohn, J. (1932) The effect of *Colpidium* on ammonia production by soil bacteria. *Annals of Applied Biology* 19, 584–608.

Morse, D.E., Hooker, N., Duncan, H. and Jensen, L. (1979) Gamma-aminobutyric acid, a neuro-transmitter, induces planktonic abalone larvae to settle and begin metamorphosis. *Science* 204, 407–410.

Nasir, S.M. (1923) Some preliminary investigations on the relationship of protozoa to soil fertility with special reference to nitrogen fixation. *Annals of Applied Biology* 10, 122–133.

Nikoljuk, V.F. (1965) Pocvennye prostejsie Uzbekistana. Izdatelstvo Akademii, Nauk, Tashkent.

Nikoljuk, V.F. (1969a) Some aspects of the study of soil protozoa. *Acta Protozoologica* 7, 99–109.

Nikoljuk, V.F. (1969b) The character of interrelationships between soil protozoa and bacteria. In: Strelkov, A.A., Sukhanova, K.M. and Raikov, I.B. (eds), *Progress in Protozoology.* Nauka, Leningrad, pp. 182–183.

Oehler, R. (1924) Weitere Mitteilungen über gereinigte Amöben- und Ciliatenzucht. *Archiv für Protistenkunde* 49, 112–134.

Old, K.M. (1977) Giant soil amoebae cause perforation of *Cochliobolus sativus. Transactions of the British Mycological Society* 68, 277–320.

Old, K.M. and Chakraborty, S. (1986) Mycophagous soil amoebae; their biology and significance in the ecology of soil-borne plant pathogens. *Progress in Protozoology* 1, 163–194.

Old, K.M. and Darbyshire, J.F. (1978) Soil fungi as food for giant amoebae. *Soil Biology and Biochemistry* 10, 93–100.

Old, K.M. and Patrick, Z.A. (1979) Giant soil amoebae: potential biocontrol agents. In: Schippers, B. and Gams, W. (eds), *Soil-borne Plant Pathogens.* Academic Press, London, pp. 617–628.

Palka, L. (1988) Rôle des protozoaires bactériophages du sol dans la minéralisation de l'azote en conditions gnotobiotiques. PhD thesis, University of Clermont-Ferrand.

Petz, W., Foissner, W. and Adam, H. (1985) Culture, food selection and growth rate in the mycophagous ciliate *Grossglockneria acuta* Foissner, 1980: first evidence of autochthonous soil ciliates. *Soil Biology and Biochemistry* 17, 871–875.

Pussard, M. (1991) Faune du sol et microflore. II – Saprophagie, prédation et médiation chimique. *Agronomie* 11, 411–422.

Pussard, M. and Rouelle, J. (1986) Prédation de la microflore. Effet des protozoaires sur la dynamique de population bactérienne. *Protistologica* 22, 105–110.

Pussard, M., Alabouvette, C. and Pons, R. (1979) Etude préliminaire d'une amibe mycophage, *Thecamoeba granifera* s.sp. minor (Thecamoebidae, Amoebida). *Protistologica* 15, 139–149.

Pussard, M., Alabouvette, C., Lemaitre, I. and Pons, R. (1980) Une nouvelle amibe mycophage endogée *Cashia mycophaga* n.sp. (Hartmannellidae, Amoebida). *Protistologica* 16, 443–451.

Ramirez, C. and Alexander, M. (1980) Evidence suggesting protozoan predation on *Rhizobium* associated with germinating seeds and in the rhizosphere of beans (*Phaseolus vulgaris* L.). *Applied and Environmental Microbiology* 40, 492–499.

Rebandel, H. and Karpinska, A. (1981) Toxic action of colistin and penicillin V and G on *Tetrahymena.* II – Inhibition of phagocytic activity. *Acta Protozoologica* 20, 291–298.

Russell, E.J. and Hutchinson, H.B. (1909) The effect of partial sterilisation of soil on the production of plant food. *Journal of Agricultural Science, Cambridge* 3, 111–114.

Sambanis, A. and Fredrickson, A.G. (1988) Persistence of bacteria in the presence of viable, non-encysting, bacteriovorous ciliates. *Microbial Ecology* 16, 197–211.

Sambanis, A., Pavlou, S. and Fredrickson, A.G. (1987) Coexistence of bacteria and feeding ciliates. Growth of bacteria on autochthonous substrates as a stabilizing factor for coexistence. *Biotechnology and Bioengineering* 29, 714–728.

Seto, M. and Tazaki, T. (1971) Carbon dynamics in the food chain system of glucose–*Escherichia coli*–*Tetrahymena vorax*. *Japanese Journal of Ecology* 21, 179–188.

Severtzova, L.B. (1928) The food requirement of soil amoebae with reference to their interrelation with soil bacteria and soil fungi. *Centralblatt für Bakteriologie* 73, 162–179.

Shikano, S., Luckinbill, L.S. and Kurihara, Y. (1990) Changes of traits in a bacterial population with protozoal predation. *Microbial Ecology* 20, 75–84.

Sinclair, J.L., McClellan, J.F. and Coleman, D.C. (1981) Nitrogen mineralization by *Acanthamoeba polyphaga* in grazed *Pseudomonas paucimobilis* populations. *Applied and Environmental Microbiology* 42, 667–671.

Singh, B.N. (1941) The influence of different bacterial food supplies on the rate of reproduction in *Colpoda steinii* and factors influencing encystation. *Annals of Applied Biology* 28, 65–73.

Singh, B.N. (1942) Selection of bacterial food by soil flagellates and amoebae. *Annals of Applied Biology* 29, 18–22.

Singh, B.N. (1945) The selection of bacterial food by soil amoebae and the toxic effects of bacterial pigments and other products on soil protozoa. *British Journal of Experimental Pathology* 26, 316–325.

Singh, B.N. (1946a) A method of estimating the numbers of soil protozoa, especially amoebae, based on their differential feeding on bacteria. *Annals of Applied Biology* 33, 112–119.

Singh, B.N. (1946b) Silica jelly as substrate for counting holozoic protozoa. *Nature (London)* 157, 302.

Singh, B.N. (1948) Studies on giant amoeboid organisms. I – The distribution of *Leptomyxa reticulata* Goodey in soils of Great Britain and the effect of bacterial food on growth and cyst formation. *Journal of General Microbiology* 2, 8–14.

Singh, B.N., Mathew, S. and Anand, N. (1958) The role of *Aerobacter* sp., *Escherichia coli* and certain amino-acids in the excystment of *Schizopyrenus russelli*. *Journal of General Microbiology* 19, 104–111.

Skinner, A.R., Anand, C.M., Malic, A. and Kurtz, J.B. (1983) Acanthamoeba and environmental spread of *Legionella pneumophila*. *Lancet* i, 289–290.

Slobodkin, L.B. (1968) How to be a predator. *American Zoologist* 8, 43–51.

Steinberg, C., Faurie, G., Zegerman, M. and Pavé, A. (1987) Régulation par les protozoaires d'une population bactérienne introduite dans le sol. Modélisation mathématique de la relation prédateur–proie. *Revue d'Écologie et de Biologie du Sol* 24, 49–62.

Stout, J.D. (1973) The relationships between protozoan populations and biological activity in soils. *American Zoologist* 13, 193–201.

Straskrabova-Prokesova, V. and Legner, M. (1966) Interrelations between bacteria and protozoa during glucose oxidation in water. *Internationale Revue der gesamten Hydrobiologie* 51, 279–293.

Taylor, W.D. (1986) The effect of grazing by a ciliated protozoa on phosphorus limitation of heterotrophic bacteria in batch culture. *Journal of Protozoology* 33, 47–52.

Telegdy-Kovats, L. de (1932) The growth and respiration of bacteria in sand cultures in the presence and absence of protozoa. *Annals of Applied Biology* 19, 65–86.

Thomas, J.D. (1990) Mutualistic interactions in freshwater modular systems with molluscan components. *Advances in Ecological Research* 20, 125–178.

Wilkinson, J.F. (1958) The extracellular polysaccharides of bacteria. *Bacteriological Reviews* 22, 46–73.

Woods, L.E., Cole, C.V., Elliott, E.T., Anderson, R.V. and Coleman, D.C. (1982) Nitrogen transformations in soil as affected by bacterial–microfaunal interactions. *Soil Biology and Biochemistry* 14, 93–98.

Wright, S.J.L., Forrest, H.S., Redhead, K. and Varnals, C.W. (1978) Predation of blue-green algae by *Acanthamoeba castellanii. Journal of Protozoology* 25, 11B.

Wright, S.J.L., Redhead, K. and Maudsley, H. (1981) *Acanthamoeba castellanii,* a predator of cyanobacteria. *Journal of General Microbiology* 125, 293–300.

Zwart, K.B. and Darbyshire, J.F. (1992) Growth and nitrogenous excretion of a common soil flagellate, *Spumella* sp. – a laboratory experiment. *Journal of Soil Science* 43, 145–151.

# Soil Protozoa as Bioindicators in Ecosystems under Human Influence

**7**

W. FOISSNER

*Universität Salzburg, Institut für Zoologie, Hellbrunnerstrasse
34, A–5020 Salzburg, Austria.*

## Introduction

Protozoa are increasingly used as bioindicators in soil ecosystems. Some
150 earlier papers on this subject have been reviewed by Viswanath and
Pillai (1968) and Foissner (1987a, b, 1991). The present review summarizes
references mainly after 1985. Other applied aspects, such as pest management and bioindication in natural ecosystems by soil protozoa are extensively discussed in Old and Chakraborty (1986), Foissner (1987a, 1987b,
1991), Klein (1988), Lousier and Bamforth (1990) and in other chapters of
this book.

## Heuristic Background

Bioindicators are, in a broad ecological sense, organisms that can be used
for the detection and quantitative characterization of a certain environmental
factor or of a complex of environmental factors; a narrower definition
confines bioindicators to human influences (Bick, 1982). Several unique
features favour the use of heterotrophic soil protozoa as bioindicators
(Foissner, 1987b):

1.  Protozoa are an essential component of soil ecosystems, because of
    their large standing crop and production. Changes in their dynamics
    and community structure very probably influence the rate and kind of
    soil formation and soil fertility.
2.  Protozoa, with their rapid growth and delicate external membranes, can

react more quickly to environmental changes than any other eukaryotic organism and can thus serve as an early warning system.

3. The eukaryotic genome of the protozoa is similar to that of the metazoa. Their reactions to environmental changes can thus be related to higher organisms more convincingly than those of prokaryotes.

4. Protozoa inhabit and are particularly abundant in those soil ecosystems that almost or entirely lack higher organisms due to extreme environmental conditions, e.g. alpine regions above the timberline, Arctic and Antarctic biotopes.

5. Protozoa are not readily dislodged in soil (Kuikman *et al.*, 1990). Many (but not all!) are ubiquitous and are useful in comparing results from different regions. Differences in patterns of distribution are almost entirely restricted to passive vertical displacement; thus, the difficult problem, especially with the epigaeon, of horizontal migration does not affect the investigations.

There are, however, several factors that have apparently restricted the use of soil protozoa and even metazoa as bioindicators (Aescht and Foissner, 1992b):

1. The immense number of species; more than 1000 may occur in a square metre of forest soil. Many specialists are needed for identification and each species has specific requirements that are often incompletely known.

2. Enumeration of soil organisms is difficult and time-consuming.

3. Animals need other organisms for food. Thus, the constellation of factors is more complicated than in plants and bioindication often remains unspecific, i.e. different factors induce similar reactions.

4. Most soil organisms are inconspicuous and invisible to the naked eye, making them unattractive to many potential investigators.

Anthropogenic stress commonly leads to changes in ecosystems that are regressive and are best described as 'impoverishment'. Most perturbations are chronic and increase the abundance of opportunistic species. Such simplified communities are not capable of the same responses to stress as more complex communities. This may be either a result of fewer redundant species in the species pool capable of exploiting changing conditions, or of biological differences in the taxa found in early successional communities as compared to those taxa found in later successional and more mature stages. The causes of these differences, the underlying biology of which is poorly known, may be the inability of communities to disperse propagules to new habitats, to respond to toxic chemicals and to exclude invaders in the case of simple communities. Continual erosion of biological diversity may result in the loss of key species that regulate numbers of other taxa and allocation of resources to biomass (Cairns and Pratt, 1990).

# Methodological Tools and Problems

The successful use of protozoa and other organisms as bioindicators depends on several methodological prerequisites. Unfortunately, some basic requirements are often neglected and must thus be discussed in some detail. In fact, methodological inconveniences will be frequently mentioned and have influenced the entire review.

## Experimental design and statistics

Unfortunately, many soil protozoological studies lack a clear experimental design and appropriate statistics. Both are, however, essential for a correct interpretation of the data. In my experience, randomized blocks with at least four (better six or more) replications should be used whenever possible (e.g. Petz and Foissner, 1989b). This experimental design provides a firm basis for parametric and non-parametric statistical tests, such as analysis of variance and linear regression (cf. Köhler *et al.*, 1984).

## Estimation of species richness and individual abundance

Natural and anthropogenic factors are often assessed by studying species richness and species composition. This presupposes that species are correctly identified. Unfortunately, misidentifications are frequent and this may explain many of the conflicting results in soil protozoology (Foissner, 1987a, 1991). Furthermore, the list of species should be comprehensive whenever possible. This is not easy with soil protozoa, because most cannot be extracted directly from the soil. In many studies, enumeration has involved various culture techniques (see Foissner, 1987a, for detailed description of methods and difficulties).

Population densities are widely used in applied soil protozoology. Foissner (1987a) discussed the problems relating to the commonly used counting methods and recommended direct counts rather than the culture techniques, e.g. Singh (1946), because the latter cannot reliably estimate the abundances of the *active* protozoa. In his opinion, the culture techniques yield, at best, a rough estimation of the abundances of the active *and* cystic protozoa (but see Perey, 1925). Unfortunately, the evidence presented in 1987 is still being widely ignored. I have thus included new results of a comparative study showing convincingly that culture methods greatly overestimate the abundance of the active protozoa (Table 7.1). They do show, however, the often surprisingly high numbers of cystic protozoa (up to 50,000 ciliate cysts in 1 g of soil; Table 7.1). My hypothesis that most

**Table 7.1.** Abundance of active ciliates g$^{-1}$ dry mass of soil in the upper 0–5 cm layer of some grasslands (from Berthold and Foissner, 1993).

| Sample | Direct counts in soil suspensions[a] | Culture method[b] Active | Active + cystic |
|---|---|---|---|
| 1 | 0 | 37000 | 46000 |
| 2 | 98 | 27000 | 27200 |
| 3 | 47 | 9200 | 9400 |

[a] Method of Lüftenegger *et al.* (1988). Recovery experiments showed that abundances are underestimated by about 30%.
[b] Method of Singh (1946). The numbers of active ciliates are determined by treating part of the sample overnight with 2% HCl as recommended by Singh. The number from the acid-treated portion of the sample, i.e. cysts, is subtracted from the total number of organisms to give the number of active ciliates. The estimation of *active* ciliates is therefore *indirect* and depends on the assumption that *all* cysts survive the acid treatment and that *all* cysts will excyst in the culture medium used (diluted soil extract). This assumption is obviously incorrect; most cysts did not excyst after the acid treatment and the other procedures (soil must be repeatedly washed to remove the acid) involved in this method.

soil protozoa are inactive (cystic) is supported by recent data provided by Darbyshire *et al.* (1989) and Glasbey *et al.* (1991) suggesting that few active ciliates reside inside soil aggregates of 2–5 mm diameter.

*Data assessment*

Commonly used indicators for environmental distress are nutrient imbalance (decrease in some nutrients and increase in others), reduced diversity of species, replacement of longer-lived by shorter-lived species (adapted to transitory novel environments), replacement of larger by smaller life forms, decline of biomass, and increase in population fluctuations of key species (Schwerdtfeger, 1975; Hill and Kevan, 1985). The evaluation and assessment of such data is a complex problem and discussion is restricted here to a few main points. Further information can be found in textbooks and the literature cited.

*The faunal approach*

Weigmann (1987) listed five criteria for assessing stress on ecosystems using species richness and species composition of the animals present.

1. Species richness in disturbed ecosystems should be compared with undisturbed ecosystems of the same ecotype. It is usually unsatisfactory, for example, to compare species richness and species composition before and after a forest is transformed into a meadow.

2. A faunal census should take account of whether a certain perturbation lowers or increases habitat diversity. Thus, increased species richness can indicate not only an improvement, but also a disturbance of an ecosystem.

3. Only taxa that have a high species richness in undisturbed ecosystems of the type being assessed should be used as bioindicators. Usually, only species-rich groups have sufficient ecological diversity to indicate a wide range of possible effects.

4. The choice of organisms used as bioindicators depends on the ecosystem and environmental hazard. This is an important, but often neglected point. Earthworms, for instance, are almost absent in strongly acidic spruce forests and above the timberline. In such ecosystems, other groups, e.g. the dominant acidophilic testate amoebae, should be used to assess certain influences. Similarly, the relative weighting of dominant to rare and/or stenoecious species depends on the specific problem under investigation.

5. Isovalent groups of indicator species can usually assess only a single disturbance factor. Amelioration of a moorland, for instance, can increase total species richness, but if a certain valency group completely disappears the amelioration must be assessed negatively.

## Coenotic indices

Many coenotic indices have been suggested to reduce the complexity of biological communities to a simple, comprehensible number or value (Washington, 1984). However, all indices reflect only few of the basic community parameters (mostly species number and abundance) and many do not consider practical needs or the most common structure of biological coenoses. A typical example is the well-known Shannon–Wiener 'diversity index', which combines in a single value individual abundance, species richness and dominance. However, dominance is misinterpreted by this index, i.e. the diversity becomes highest in such communities where all species have the same abundance. Such coenoses hardly ever exist, because most distribution patterns fit a logarithmic series (Magurran, 1988).

Wodarz *et al.* (1992) have thus suggested a 'Weighted Coenotic Index' (WCI) that unifies in a single value the total abundance and the logarithmic dominance structure as well as species richness and ecological weightings, e.g. habitat preferences and positions of species in the $r/K$-continuum.

$$\text{WCI} = \sum_{i=1}^{S} 10^8 \cdot \underbrace{\frac{p_i \cdot (1-p_i)^{S-1}}{[o_i \cdot (o_{max}-o_i)+1]}}_{\substack{\text{distribution} \\ \text{component}}} \times \underbrace{\frac{n_{max} \cdot \tilde{n}_i}{[(n_{max}-\tilde{n}_i) \cdot (\tilde{n}_i-n_{min})+1]}}_{\substack{\text{distribution} \\ \text{constant}}} \times \underbrace{\frac{w_{1i} \cdot w_{2i} \cdot w_{ni}}{N \cdot S}}_{\substack{\text{abundances and} \\ \text{weightings}}} \cdot \frac{1}{S}$$

$$\text{for } p_i = S \neq 1$$

where:

| | |
|---|---|
| $n_i$ | = number of individuals in species i |
| $\tilde{n}_i$ | = median of the $n_i$ values |
| $n_{max}$ | = the highest $n_i$ value in a sample |
| $n_{min}$ | = the lowest $n_i$ value in a sample |
| $N$ | = the total number of individuals |
| $o_i$ | = $1 + \text{ld}(n_i)$ [ ld = logarithmic dominance structure] = the octave species i belongs to |
| $o_{max}$ | = $1 + \text{ld}(n_{max})$ = the highest octave in a sample |
| $p_i$ | = $n_i/N$ = the relative abundance of species i |
| $S$ | = species richness, i.e. total number of species in all replicates |
| $w_{1i}$ | = weighting 1, e.g. degree of autochthonism in species i |
| $w_{2i}$ | = weighting 2, e.g. the position of species i in the $r\,K$ continuum |
| $w_{ni}$ | = further weightings |
| $10^8$ | = factor used to transform values to integers |

Studies with simulated biocoenoses showed that ecological weighting and dominance structure are major components of the index; the ecological weighting needs to be related to the group of organisms studied and to the scope of the investigation. The WCI is a relative measure that needs a reference (control) site for a conclusive interpretation. Compared to several other diversity indices, WCI is an improvement, because of the inclusion of ecological weighting and the logarithmic dominance structure. This index was applied to published data from several field studies using protozoa (testate amoebae, ciliates) and earthworms. The results show that this index is an appropriate measure of changes and recovery processes in disturbed communities (Tables 7.2 and 7.19). The WCI is certainly no substitute for the classic methods (e.g. Schwerdtfeger, 1975; Southwood, 1978) of describing animal communities, but is very useful in summarizing data.

## Scaling and persistency of disturbances

Reactions of organisms to detrimental influences have been extensively discussed by Domsch *et al.* (1983) and Beck *et al.* (1988). Doubling time and reproductive potential of the populations and sociopolitical constraints determine scaling and assessment of the reactions. Domsch *et al.* (1983) suggested that side-effects persisting for more than 60 days may be critical to populations of single-celled organisms.

**Table 7.2.** Weighted Coenotic Index (WCI) and Shannon–Wiener diversity (H') of soil testacean communities. *N*, total abundance (individuals g⁻¹ dry mass of soil); *S*, species number. (From Wodarz *et al.*, 1992.)

| Sites and treatments (designated according to the original papers) | Testaceans[a] | | | |
|---|---|---|---|---|
| | *N* | *S* | WCI | H' |
| Agricultural soils[b] | | | | |
| Organically farmed | 884 | 28 | 33 | 2.896 |
| Conventionally farmed | 528 | 25 | 111 | 2.898 |
| Vineyard soils[c] | | | | |
| Minimal (reference site) | 270 | 13 | 2285 | 1.279 |
| Conventional | 156 | 12 | 46123 | 1.335 |
| Biodynamic | 239 | 14 | 18730 | 1.552 |
| Organic-biological | 347 | 16 | 2181 | 1.557 |
| Semi-biological | 748 | 21 | 162 | 1.706 |
| Soil compaction experiment[d] | | | | |
| Control (chamber effect) | 1816 | 11 | 36 | 1.489 |
| Soil compaction 10% | 1709 | 10 | 248 | 1.924 |
| Soil compaction 30% | 906 | 8 | 973 | 1.802 |
| Soil compaction 50% | 151 | 5 | 263719 | 1.519 |

[a] Decreasing WCI values indicate improving soil conditions.
[b] Original data from Foissner (1992) and unpublished.
[c, d] Original data from Lüftenegger and Foissner (1989a) and Berger *et al.* (1985) respectively. The WCI differentiates the treatments more clearly than the Shannon–Wiener index. Furthermore, the detrimental effect of, for example, a 50% soil compaction is much better expressed by the WCI than by H'.

# Soil Protozoa as Bioindicators in Ecosystems under Human Influence

## *Effects of irrigation*

Data from field and laboratory studies of the effects of irrigation on abundance and species composition of soil protozoa are contradictory (Foissner, 1987a). Two recent field studies have also produced conflicting results. Szabó (1986) stated that the total numbers of protozoa and the abundances of the ciliates were strongly related to the water content in a Hungarian chernozem soil under maize cultivation. Flagellates and naked amoebae were active at water contents as low as 11–18% vol, whereas ciliates usually encysted if soil moisture was less than 13% vol. Unfortunately, Szabó (1986) omitted to mention either the method used for counting the protozoa (probably a culture method) or whether the counts relate only to active or to

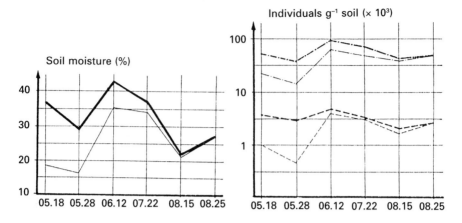

**Fig. 7.1.** Effects of irrigation on the moisture content and the protozoan numbers in a chernozem soil from May to August. —— irrigated, —— control, — · — · total number of protozoa in irrigated soil, — · — · number of ciliates in irrigated soil, — — — total number of protozoa in control soil, — — — number of ciliates in control soil. (From Szabó, 1986.)

both active and cystic individuals. Furthermore, the data were not evaluated statistically. At first glance, Fig. 7.1 seems to support Szabó's (1986) conclusions. A closer inspection shows, however, that the protozoa retained a much higher abundance in the irrigated soil than in the non-irrigated control, although the water contents of the treatments were very similar during the last three periods of investigation.

Petz and Foissner (1989a) observed a marked decrease ($P \leq 0.05$) in the abundance of the active ciliates in the litter layer of an irrigated spruce forest stand (Table 7.3); species richness, however, increased in both ciliates and testate amoebae, indicating that certain species need higher soil moisture contents. Testacean numbers did not increase significantly in response to irrigation. The nematode numbers increased by about 45% in the 0–3 cm layer of the same irrigated plots. Lousier (1974a,b), however, found significant positive correlations between soil moisture and number of total and active testaceans in a very dry aspen woodland after irrigation ($r = 0.873$ and 0.804, respectively). Wanner and Funke (1989) and Wanner (1991), like Petz and Foissner (1989a), did not find a significant correlation between soil moisture and testacean numbers (Tables 7.3 and 7.5).

In Antarctic biotopes, Smith (1973, 1985) found significant positive correlations between the flagellate *Cercobodo vibrans*, the testate amoeba *Corythion dubium* and the moisture contents of both bare fellfield fines and *Andreaea* moss cushions; correlation coefficients with maximum soil temperature were smaller. Smith (1985) therefore suggested that moisture

**Table 7.3.** Effects of irrigation on the microfauna of a spruce forest soil[a] (from Petz and Foissner, 1989a).

| Parameters | Soil depth (cm) | Irrigated[b] | Control | n | Statistics |
|---|---|---|---|---|---|
| Soil moisture (% wet mass of air-dried soil) | 0–3 | 50.0* ± 9.9 | 41.3† ± 12.6 | 15 | ANOVA ($P \leq 0.05$) |
|  | 3–9 | 48.8* ± 4.3 | 44.7* ± 5.9 | 8 | ANOVA ($0.2 \geq P > 0.1$) |
| Ciliates |  |  |  |  |  |
| abundance | 0–3 | 311* ± 141 | 489† ± 258 | 15 | U-test ($0.05 < P \leq 0.1$) |
|  | 3–9 | 10* ± 10 | 14* ± 24 | 8 | U-test ($P \geq 0.2$) |
| species number | 0–3 | 12.8* ± 4.5 | 8.4† ± 3.3 | 15 | ANOVA ($P \leq 0.005$) |
|  | 3–9 | 2.4* ± 2.1 | 1.6* ± 1.9 | 8 | U-test ($P \geq 0.2$) |
| Testaceans |  |  |  |  |  |
| abundance |  |  |  |  |  |
| living | 0–3 | 22203* ± 7040 | 17908* ± 3175 | 4 | ANOVA ($P \geq 0.2$) |
| empty tests | 0–3 | 361720* ± 17204 | 319081* ± 58390 | 4 | ANOVA ($P \geq 0.2$) |
| species number |  |  |  |  |  |
| living | 0–3 | 11.5* ± 1.3 | 9.8† ± 1.0 | 4 | ANOVA ($0.05 < P \leq 0.1$) |
| total | 0–3 | 18.0* ± 0 | 19.8* ± 2.2 | 4 | U-test ($P \geq 0.2$) |
| Nematodes |  |  |  |  |  |
| abundance | 0–3 | 1197* ± 349 | 824† ± 349 | 15 | ANOVA ($P \leq 0.01$) |
|  | 3–9 | 308* ± 80 | 391* ± 173 | 8 | ANOVA ($P \geq 0.2$) |
| Rotifers |  |  |  |  |  |
| abundance | 0–3 | 227* ± 88 | 181* ± 122 | 15 | ANOVA ($P \geq 0.2$) |
|  | 3–9 | 38* ± 25 | 31* ± 18 | 8 | U-test ($P \geq 0.2$) |

[a] Abundances (individuals $g^{-1}$ dry mass of soil; arithmetic mean ± standard deviation) were estimated with the direct counting method of Lüftenegger et al. (1988). Values followed by the same symbol are not significantly different.
[b] Irrigated plot (15 m²) received 25 l m⁻² water every fourth day.

rather than temperature is the principal environmental factor limiting the rate of development of biotic communities in Antarctic fellfields. Other studies, however, did not find any correlation between flagellate abundance and soil moisture (for review see Foissner, 1987a).

The data available (Foissner, 1987a, and this review) suggest that soil protozoa do not depend as much on moisture as was widely assumed and often concluded from microcosm experiments (e.g. Kuikman *et al.*, 1991). It seems likely that moderate, artificial irrigation often does not markedly influence soil protozoa. Irrigation or reclamation probably increases individual abundance and species richness only in very dry (e.g. deserts and semideserts) or very moist biotopes. Many protozoa are obviously active in rather dry soils and can live in very thin films of water. In the experiment of Petz and Foissner (1989a), the litter of the control plot appeared virtually dry; most of the water present was a constituent of the needles. Foster and Dormaar (1991) showed in an ultrastructural study that amoebae can probably exploit micropores having a diameter of only 1 μm.

## Effects of organic and mineral fertilizers

Most of the extensive literature available relating to the effects of soil fertilizers on protozoa is reviewed in Foissner (1987a). The few, more recent papers discussed below support the general conclusion that most fertilizers increase the numbers of protozoan cells and change the dominance structure of the soil protozoa.

Paustian *et al.* (1990) studied the effects of nitrogen amendments on the carbon and nitrogen budgets and the soil fauna in four Swedish agroecosystems with annual and perennial crops (Fig. 7.2). The variation in the N input (1–39 g N m$^{-2}$ year$^{-1}$) and cropping system influenced primary production (260–790 g C m$^{-2}$ year$^{-1}$) and the input of organic material to the soil (150–270 g C m$^{-2}$ year$^{-1}$). This was reflected in variations of total soil animal biomass (1.6–5.1 g C m$^{-1}$) and in variations in the abundance of the nematodes and the micro- and macroarthropods. In contrast, total bacteria, fungi, flagellates and amoebae varied quite independently of the organic matter input (Sohlenius, 1990; Fig. 7.2a-c). Protozoa showed few differences between treatments and most amoebae occurred in the unfertilized plot (Fig. 7.2b). The average number of amoebae was significantly higher ($P \leq 0.05$) in GL (meadow fescue ley receiving 120+80 kg N ha$^{-1}$ year$^{-1}$) than in B120 (barley receiving 120 kg N ha$^{-1}$ year$^{1}$). Schnürer *et al.* (1986) speculated that the difference in the input of organic matter was too small to produce pronounced effects, whereas Sohlenius (1990) suggested that the rather constant microbial biomass was a result of an adjustment in the grazing pressure of microbial-feeding animals to the level of microbial production. However, methodological shortcomings of the culture method,

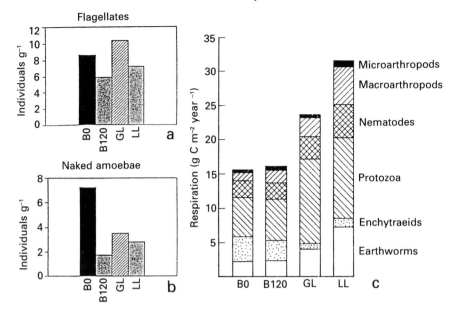

**Fig. 7.2a – c.** Mean numbers of flagellates and naked amoebae (g⁻¹ dry mass of soil), and mean annual rates of respiration for the total fauna (herbage and soil) in barley without N fertilizer (BO), barley with 120 kg N ha⁻¹ year⁻¹ (B120), grass ley with 200 kg N ha⁻¹ (GL), and lucerne (LL). A randomized block design with four replicates was used. Total (active + cystic) abundances of flagellates and naked amoebae were estimated by a culture method in 20 samplings over three years. (From Paustian *et al.*, 1990, and Sohlenius, 1990.)

where cystic and active protozoan cells were not distinguished, could also have contributed to this rather unexpected result.

Gupta and Germida (1988) investigated the effects of five years of repeated application of elemental S fertilizer on the soil protozoa in S-deficient soils in western Canada. The application of S⁰ fertilizer reduced the microbial biomass and its activity in soil. Soils treated with 44 kg S⁰ ha⁻¹ year⁻¹ for five years exhibited a 30–71% decline in protozoa feeding on bacteria and more than a 84% decline in the population of mycophagous amoebae. This decline in protozoan populations accompanied changes in microbial biomass, especially in the case of mycophagous amoebae and fungal biomass (Fig. 7.3). The adverse effect of repeated S⁰ applications on microbial biomass and predatory protozoa was persistent. Since nutrient transformations (e.g. mineralization) in soil are influenced by microbial interactions, these results suggest reduced nutrient turnover via microbial predation in S⁰-treated soils.

Tirjaková (1991) investigated the effects of some agricultural fertilizers

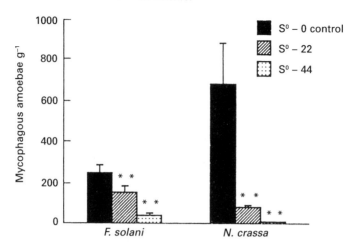

**Fig. 7.3.** Effect of 5 years of elemental S fertilizer application on the mycophagous soil amoebae feeding on *Fusarium solani* or *Neurospora crassa*. $S^0$ applied at 22 or 44 kg ha$^{-1}$ yr$^{-1}$. Triplicate samples from brome grass–alfalfa pasture sites were analysed. Amoebae were counted with Singh's ring technique (1946) using a most probable number method. ** Significant at $P \leq 0.001$. (From Gupta and Germida, 1988.)

(NPK, urea, ammonium–calcium nitrate, ammonium sulfate) on ciliates in soil microcosms. The numbers of species decreased with increasing fertilizer concentration (2, 10, 20, 50 g l$^{-1}$); the type of fertilizer was of minor importance. Most species occurred rather frequently up to a concentration of 10 g l$^{-1}$; 20 g l$^{-1}$ were toxic for most species. Only seven species (compared to 71 species found in the control) survived 50 g l$^{-1}$, viz. *Colpoda aspera*, *C. steinii*, *Enchelys gasterosteus* (probably a misidentification as this species very likely does not occur in soils), *Hemisincirra gellerti*, *Homalogastra setosa*, *Leptopharynx costatus* and *Platyophrya spumacola*. Of the four fertilizers tested, NPK was the least toxic.

Aescht and Foissner (1991, 1992) investigated protozoa (testate amoebae, ciliates), small metazoa (nematodes, rotifers) and soil enzymes (catalase, cellulase) in a reforested fertilized site at the alpine timberline (Fig. 7.4). None of the treatments caused a significant decrease of the biological parameters investigated in comparison with untreated controls. Soil life was more or less stimulated depending on the quantity and organic content of the fertilizers; 180–270 g organic material per seedling were found to be most effective. Dried bacterial biomass increased the pH by about 0.5 units, the catalase activity by about 70%, and the number of ciliates and nematodes by 150–400%. Biomass and species number of ciliates were likewise

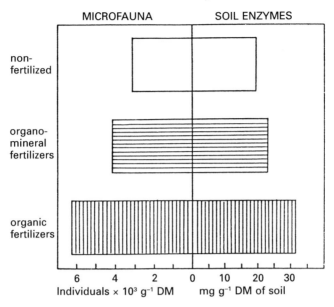

MICROFAUNA          SOIL ENZYMES

non-
fertilized

organo-
mineral
fertilizers

organic
fertilizers

```
   6    4    2    0   10   20   30
Individuals × 10³ g⁻¹ DM   mg g⁻¹ DM of soil
```

**Fig. 7.4.** Mean numbers of soil animals (testate amoebae, ciliates, nematodes, rotifers) and enzymatic activites (catalase, cellulase) in a reforested and fertilized site near the alpine timberline. A randomized block design with six replicates for four years was used. Mineral and organic fertilizers were applied separately (90 g NPK; 90, 180, 300, 450 g dried bacterial biomass per spruce seedling, respectively) and in combination with magnesite (90 g NPK + 300 g Mg; 90, 180, 300 g bacterial biomass + 300 g Mg each, respectively; 30 g dried fungal biomass + 270 g Mg). These fertilizers were applied in a granular form within a 10 cm radius around each seedling. Testate amoebae, nematodes and rotifers were counted with the direct method of Lüftenegger *et al.* (1988). The 'potential' abundance of the ciliates was estimated with a culture method. (From Aescht and Foissner, 1991, 1992.)

increased. Organo-mineral fertilizers caused a pH rise of up to two units and also stimulated soil life; but efficiency decreased markedly if the organic content was less than 180 g per seedling. The lowest biological activities were observed in the control and the soil fertilized with NPK. Testaceans, rotifers and cellulolytic activities were not significantly affected by the treatments. Pooled evaluation of the data (organic versus organo-mineral treatments) and community analyses show that the organic fertilizers caused a more pronounced increase in soil life and changes in the community structure than the mineral combinations (Fig. 7.5). Two years after refertilization, the differences between treatments and unfertilized controls had diminished.

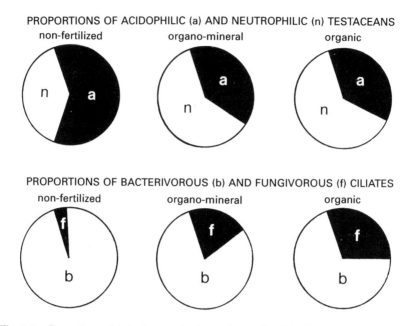

PROPORTIONS OF ACIDOPHILIC (a) AND NEUTROPHILIC (n) TESTACEANS

non-fertilized                organo-mineral                organic

PROPORTIONS OF BACTERIVOROUS (b) AND FUNGIVOROUS (f) CILIATES

non-fertilized                organo-mineral                organic

**Fig. 7.5.** Proportions of indicator species in a reforested and fertilized site near the alpine timberline. See Fig. 7.4 for methods. (From Aescht and Foissner, 1991, 1992.)

## Effects of forest liming and fertilizers

A dramatic forest decline is evident in regions exposed to heavy air-pollution. The decline is usually associated with a significant decrease in soil pH and/or essential plant nutrients. Thus, liming and fertilizers are widely used to overcome the problem. However, many forest ecologists and soil zoologists are disquieted by such treatments, as they might induce unexpected long-term changes (Funke, 1987). As regards soil protozoa, the data reviewed in Foissner (1987a) indicate that liming and fertilizers increase the individual and species numbers of the ciliates and testate amoebae in very acidic (pH ≤ 4) forest soils. Ciliates react particularly quickly to changes in the physico-chemical and biological properties of their environment. Liming increases the number of species and individuals, but fertilizers decrease both. If liming is combined with fertilizers, the positive and negative effects seem to counteract each other (Lehle and Funke, 1989; Funke and Roth-Holzapfel, 1991). More detailed investigations using randomized blocks showed, however, that treatment effects are temporary and not very pronounced (Rosa, 1974; Wanner and Funke, 1989; Wanner, 1991; Aescht

**Table 7.4.** Average individual number of protozoans in fertilized spruce forest plots.

| Treatments | Flagellates | Ciliates | Testaceans | Total |
|---|---|---|---|---|
| Control[a] | 5323 | 1650 | 1534 | 8507 |
| NPKCa[b] | 7569 | 3051 | 1319 | 11939 |
| N[c] | 7031 | 2414 | 1723 | 11168 |
| Old stand[d] | | | | |
| Control | 9553 | 583 | 32026 | 42162 |
| Biomag[e] | 5403 | 361 | 38734 | 44498 |
| Bactosol + Biomag[f] | 9634 | 453 | 34660 | 44747 |
| Young stand[d] | | | | |
| Control | 1696 | 286 | 39944 | 41926 |
| Biomag[e] | 2263 | 229 | 34063 | 36555 |
| Bactosol + Biomag[f] | 3830[g] | 221 | 36531 | 40582 |

[a] Data from Rosa (1974) and compiled from tabulated values. A randomized block design with five replicates was used; each block was sampled 8–9 times during a period of one year. A rather dubious culture method was used: 5 g humus were added to 10 ml distilled water. After 24 hours the individuals present were counted and the number g$^{-1}$ humus calculated from two replicates.
[b] Fertilized with 200 kg N, 100 kg P, 100 kg K and 300 kg lime ha$^{-1}$.
[c] Fertilized with 100–150 kg N ha$^{-1}$.
[d] Data from Aescht and Foissner (1993 and unpublished). A randomized block design with four (testaceans) to six (ciliates and flagellates) replicates was used; blocks were sampled 7–8 times during a period of five years, except for the flagellates which were sampled once, viz. two years after fertilization. Counts were made with the direct method as described by Lüftenegger *et al.* (1988) and are given as cells g$^{-1}$ dry mass of soil.
[e] Fertilized with 1800 kg magnesite and 200 kg dried fungal biomass ha$^{-1}$.
[f] Fertilized with 3000 kg dried bacterial biomass, 1800 kg magnesite and 200 kg dried fungal biomass ha$^{-1}$.
[g] Significant difference from control (ANOVA).

and Foissner, unpublished; Tables 7.4 and 7.5). Certainly, the effects depend on the amount and kind of substances applied and may be masked by the detrimental effects of heavy metals, organics (polycyclic aromatic hydrocarbons), etc. usually deposited together with the acidifying NOx and S compounds.

Rauenbusch (1987) observed a decrease in numbers and an increased rate of malformed euglyphid testaceans in very acidic forest soils; detailed evidence is, however, not provided.

Applications of slow-acting pH-regulators and fertilizers seem unlikely to disturb soil fauna greatly and might be recommended prudently, if the vitality of the trees is increased and the groundwater does not become eutrophic from leachates.

**Table 7.5.** Effects of liming, fertilization, irrigation, acidification and pesticides on the testate amoebae in spruce forest stands near Ulm, Germany (compiled from Wanner, 1991).

| Sites, investigation periods and treatments[a] | n[b] | pH (KCl) | Testaceans[c] | | |
|---|---|---|---|---|---|
| | | | Individuals | Biomass | Taxa |
| Site A ( June 1984–Oct. 1987) | | | | | |
| Control | 9 | 3.1* | 9091* | 0.111* | 18.1* |
| Limed + fertilized | 9 | 4.5† | 6926* | 0.068* | 15.1† |
| Limed + fertilized + | | | | | |
| irrigated | 9 | 3.2‡ | 8021* | 0.083* | 14.2‡ |
| | | | | | |
| Site B (April 1984–Oct. 1987) | | | | | |
| Control | 10 | 2.9* | 7017* | 0.073* | 15.0* |
| Limed | 10 | 3.5† | 10137† | 0.122† | 15.8* |
| | | | | | |
| Site C (Nov. 1986–May 1988) | | | | | |
| Control | 4 | 3.5 | 15800 | 0.194 | 19.2 |
| NaCl | 4 | 3.5 | 12741 | 0.142 | 18.0 |
| $H_2SO_4$ | 4 | 3.0 | 13215 | 0.098 | 16.2 |
| Lindane | 4 | 3.5 | 15165 | 0.176 | 17.2 |
| Ripcord | 4 | 3.5 | 10423 | 0.110 | 16.7 |

[a] Limed + fertilized: 20 kg 95% $CaCO_3$ 100 $m^{-2}$ + 5 kg $5Ca(NO_3)_2NH_4NO_3$ 100 $m^{-2}$.
Irrigated: to –70 cm water holding capacity.
Limed: 20 kg 95% $CaCO_3$ 100 $m^{-2}$.
NaCl: 2.25 kg $m^{-2}$.
$H_2SO_4$: 10 l $m^{-2}$ (5%).
Lindane: 120 g $ha^{-1}$.
Ripcord: 10 g $m^{-2}$.
[b] Number of samples investigated.
[c] Numbers (direct counts in aqueous soil suspensions) are given as active individuals $g^{-1}$ dry mass of soil. Biomasses: mg $g^{-1}$ dry mass of soil. Values followed by the same symbol are not significantly different ($P \geq 0.05$).

## Effects of soil management

Most investigations relating to the effects of cultivation, crop rotation and organic farming on soil protozoa were reviewed in detail by Foissner (1987a). The few more recent studies that have been published agree with the previous conclusions that cultivation and organic farming tend to increase the abundance and species richness of soil protozoa compared with virgin lands and fallows (Mordkovich, 1986; Tirjaková, 1988; Foissner, 1992). Tirjaková (1988) investigated 147 samples from 22 agricultural soils in Czechoslovakia and found a characteristic community of ciliates: *Colpoda inflata, C. steinii, Cyrtolophosis elongata; Gonostomum affine, Histriculus muscorum, Homalogastra setosa, Leptopharynx costatus*; the proportion of rare species was high.

Foissner *et al.* (1986, 1990), Foissner (1987c) and Lüftenegger and Foissner (1989a) studied some ecofarmed and conventionally farmed fields and grasslands in Austria with special reference to the protozoa. The results obtained from the evaluation of a total of 13 paired sites (ecofarmed vs. conventionally farmed) have been summarized by Foissner (1992):

1.  Many of the soil zoological parameters under investigation did not differ statistically in ecofarmed and conventionally farmed fields and grasslands.
2.  There were no striking differences in species composition and dominance structure of the ciliates and testate amoebae.
3.  All differences that can be guaranteed with an error probability of $\alpha = 10\%$ or less invariably show higher biological activity in the ecofarmed plots (Fig. 7.6). The soil physical and chemical investigations which accompanied the zoological studies of some sites revealed larger biological activity is correlated with the larger humus content and smaller soil compaction. The organic matter content is significantly larger in the

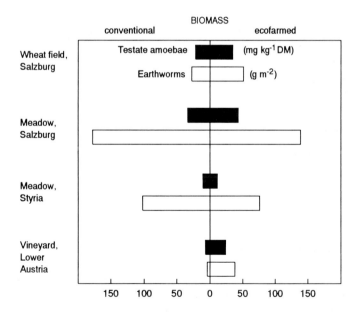

**Fig. 7.6.** Biomass of testate amoebae (direct counts) and earthworms (formaldehyde extraction) in conventionally farmed and ecofarmed fields and grasslands in Austria. Combined data from several investigations. DM, dry mass of soil. (From Aescht and Foissner, 1991.)

ecofarmed plots, whereas soil compaction is more pronounced under conventional cultivation.

4. Conventional agriculture has a more detrimental effect on soil fauna in semiarid regions without animal husbandry than in Atlantic regions with mixed farming.

These and other results from the literature show that generalizations like 'conventional farming destroys life in the soil' or 'ecofarming stimulates soil life' are only partially supported by the available data. A far more comprehensive analysis including climate, soil type and farm management is required. However, the detrimental effects caused by conventional farming already discernible on soil fauna call for serious consideration and ought to stimulate the development of more compatible agricultural technology and intensified soil biological research. Future research should include studies on productivity of soil animals under various management systems, the analysis of single factors (e.g. the special admixtures used in biodynamic farming) to elucidate causative mechanisms, and studies on the relationship between soil animals, crop production and sustained yield.

Zoologists are frequently perplexed by practical questions posed by farmers. For example, 'What do I stand to gain by having say about 10% more soil animals in my fields? Will my yield also go up by 10% or will I get a more sustainable yield?' Unfortunately, soil zoologists cannot provide satisfactory answers to these important questions, because of insufficient data related to crop yield. There is an urgent need for integrating soil zoological studies with plant agronomy and agricultural economics. Only when we can provide conventional farmers with conclusive evidence of the economic importance of soil animals, may we expect to convince them of the benefits of ecofarming.

## Effects of biocides

Biocides, including pesticides and herbicides, play a major and increasingly controversial role in modern agriculture. Foissner (1987a) reached four conclusions from the data available:

1. The general pattern of reaction of soil protozoa to biocide stress is the same as that of other organisms.
2. Many protozoan species seem to be just as sensitive to pesticides as other more commonly used test organisms.
3. Insecticides are usually more toxic than herbicides.
4. Insecticides disturb soil protozoa critically, i.e. populations often do not fully recover within 60 days.

These conclusions have been substantiated and extended by other studies. These are reviewed in some detail in this chapter since there is a strong possibility of replacing vertebrates by protozoa as test models for pesticides (Pons and Pussard, 1980). The problems of adaptation to biocide stress are, however, also evident in protozoa (Raederstorff and Rohmer, 1987a, 1987b). The systemic fungicides tridemorph and fenpropimorph decreased the growth rate of the soil amoeba, *Acanthamoeba polyphaga*, at a 40 μM concentration. After several weekly subcultures, the growth rate returned to normal and the inhibitor concentration stabilized at 80 μM. After four to five extra subcultures, the growth rate was again identical to that in untreated control cultures. The amoebae could tolerate even higher concentrations of inhibitors (up to 160 μM) without apparent damage. Their sterol biosynthesis was, however, greatly influenced. Not only was there a much larger amount of sterols in the fungicide-exposed cells, but these were also synthesized via a different metabolic pathway from those found in the control cells. These results are confirmed and extended by a recent paper of Groliere *et al.* (1992), indicating that *Tetrahymena pyriformis* GL can convert a dithiocarbamate fungicide, thiram, into metabolites toxic to this ciliate.

Another new approach to the problem is taken in the paper by Pižl (1985). He found a significantly increased infection of earthworms by monocystid gregarines when the earthworms were exposed to a herbicide, zeazin 50, for 26 weeks (Table 7.6).

Schreiber and Brink (1989) devised an *in vitro* toxicity test for pesticides using soil and freshwater protozoa as test organisms. They observed quite variable sensitivities, and the conventional soil application rates used by farmers for chlorex, MCPA, and benlate were toxic to some of the organisms (Table 7.7).

There is still a great need for well-designed field studies on the effects of biocides on soil protozoa, particularly on testate amoebae (Foissner, 1987a). The only such studies available are those by Petz and Foissner

**Table 7.6.** The effect of the herbicide zeazin 50 on the incidence of infection of earthworms by monocystid gregarines (from Pižl, 1985).

| | % incidence of infection[b] | | | |
|---|---|---|---|---|
| Species[a] | Control without cysts | Control with cysts | 5 kg ha⁻¹ zeazin 50 | 40 kg ha⁻¹ zeazin 50 |
| *Lumbricus castaneus* + | 0 | 64.0 | 83.3** | 100.0** |
| *Lumbricus terrestris* + | 0 | 56.0 | 84.0** | 96.0** |
| *Octolasium lacteum* ++ | 0 | 36.0 | 52.0* | 75.0* |

[a] Symbols: + earthworms infected by *Monocystis agilis*, ++ earthworms infected by *Aplocystis herculea*.
[b] $\chi^2$-test: *$P \leqslant 0.05$; ** $P \leqslant 0.01$.

**Table 7.7.** Conventional agricultural concentrations (CC) applied to arable land, and 9-hour lethal concentrations for six pesticides and the protozoans *Colpoda cucullus* (*C. c.*), *Blepharisma undulans* (*B. u.*) and *Oikomonas termo* (*O. t.*). (From Schreiber and Brink, 1989.)

| Pesticide | CC (ppm) | 9-h $LC_{50}$[a] (ppm) | | |
|-----------|----------|----------|----------|----------|
|           |          | *C. c.* | *B. u.* | *O. t.* |
| Chlorex | 1000 | 320 | 360 | 40000 |
| MCPA | 3 | 100 | 1 | 2 |
| Benlate | 1 | 4 | 0.7 | 10 |
| Dichlorprop | 2 | >>100 | >>100 | >>100 |
| Matrigon | 0.5 | >>100 | >100 | 4 |
| Sumicidin | 2 | >>100 | >100 | 6 |

[a] Concentrations at which 50% ($LC_{50}$) of protozoan population died after 9-hour incubation.

(1989b), Wanner and Funke (1989) and Wanner (1991). Petz and Foissner (1989b) investigated the effects of a fungicide, mancozeb and an insecticide, lindane, on the active microfauna of a spruce forest soil using a completely randomized block design and a direct counting method (Table 7.8). The effects were evaluated 1, 7, 15, 40, 65 and 90 days after application of a standard or high ($\times$ 10) dose (0.096 g $m^{-2}$ and 6 g $m^{-2}$ active ingredient, respectively). Mancozeb, even at the higher dose, had no pronounced acute or long-term effects on absolute numbers of the taxa investigated. The number of ciliate species decreased one day after treatment with the standard dose ($0.05 < P \leq 0.1$), but soon recovered (Fig. 7.7). However, the community structure of ciliate species was still slightly altered after 90 days. Mycophagous ciliates were distinctly reduced in the first weeks after application of the fungicide (Table 7.9). Testaceans were not reduced before day 15 with the higher dose or before day 40 with the standard dose ($0.05 < P \leq 0.1$). A standard dose of lindane caused acute toxicity in ciliates and rotifers ($P \leq 0.05$), although the latter soon recovered. The number and community structure of ciliate species were still distinctly altered after 90 days ($0.05 < P \leq 0.1$), indicating the crucial influence of lindane. Testaceans were reduced only on day 15 and nematodes only on day 40 ($0.05 < P \leq 0.1$). At the high dose of lindane, severe long-term effects occurred in soil moisture, total rotifers ($P \leq 0.05$), total nematodes ($0.05 < P \leq 0.1$) and in the structure of the ciliate community. Some species were encouraged by lindane after 90 days, e.g. *Colpoda inflata*, *C. steinii* and *Pseudoplatyophrya nana*, possibly due to reduced competition and their *r*-selected survival strategy, whereas *Avestina ludwigi*, very dominant in the control plots, became extinct (Table 7.8). Generally, there were marked differences

**Table 7.8.** Percentage of the dominant species of active ciliates and testaceans 1 and 90 days after treatment with mancozeb and lindane at normal (0.096 g m$^{-2}$ and 6 g m$^{-2}$ active ingredient, respectively) and high doses (0.96 g m$^{-2}$ and 60 g m$^{-2}$ active ingredient, respectively). (From Petz and Foissner, 1989b.)

| Species | Day | Control | Mancozeb | | Lindane | |
|---|---|---|---|---|---|---|
| | | | 1 × | 10 × | 1 × | 10 × |
| Ciliates | | | | | | |
| *Avestina ludwigi* | 1 | 23.5 | 44.3** | 37.4 | 15.2** | 0.0** |
| (Aescht & Foissner) | 90 | 41.3 | 45.8⁺ | 42.9 | 26.7⁺⁺ | 0.0** |
| *Platyophrya* | 1 | 19.4 | 12.7 | 21.7 | 12.1 | 0.0** |
| *spumacola* (Kahl) | 90 | 13.1 | 9.8 | 13.6 | 22.4⁺ | 4.4** |
| *Pseudoplatyophrya* | 1 | 18.5 | 10.0* | 6.0 | 15.2 | 0.0** |
| *nana* (Kahl) | 90 | 10.8 | 13.6 | 17.7 | 24.2** | 35.8** |
| *Colpoda steinii* | 1 | 4.5 | 1.4 | 4.0 | 6.1 | χ$^a$ |
| (Maupas) | 90 | 0.0 | 0.0 | 0.7 | 2.5⁺⁺ | 44.8** |
| *Colpoda inflata* | 1 | 0.1 | 1.0 | 0.5 | 0.0 | 0.0 |
| (Stokes) | 90 | 0.8 | 0.0 | 0.7 | 0.6⁺ | 9.3** |
| Testaceans | | | | | | |
| *Corythion dubium* | 1 | 48.1 | 74.6* | 63.3 | 46.0 | 51.3 |
| (Taranek) | 90 | 47.9 | 52.2 | 48.0 | 39.7 | 32.7 |
| *Trinema lineare* | 1 | 11.3 | 7.5 | 0.0 | 18.7 | 12.2 |
| (Penard) | 90 | 9.6 | 17.5 | 17.8 | 18.3** | 15.7 |
| *Schoenbornia* | 1 | 0.0 | 0.0 | 2.8 | 3.0 | 8.7 |
| *humicola* (Schönborn) | 90 | 9.6 | 9.7 | 6.4 | 6.9 | 17.0 |

[a] High value, not representative because only two individuals were found to be active.
* $0.05 < P \leqslant 0.1$; ** $P \leqslant 0.05$; differences from control.
⁺ $0.05 < P \leqslant 0.1$; ⁺⁺ $P \leqslant 0.05$; differences from high dose.

between the effects of the standard and the high dose of lindane, but not with mancozeb. Ciliates showed very pronounced changes after the pesticide applications, whereas testaceans were more resistant (Table 7.10).

Strong toxicity of lindane to soil protozoa has also been reported by others (for review see Foissner, 1987a). *In vitro*, tetrahymenid ciliates may survive rather high concentrations of this insecticide, i.e. 8–100 mg l$^{-1}$ (Komala, 1978; Dive *et al.*, 1980; Wiger, 1985). However, pronounced changes in shape and distinct inhibition of the synthesis of DNA, RNA and proteins occur at much lower concentrations, i.e. 2.5 mg l$^{-1}$ (Wiger, 1985; Mathur and Saxena, 1986, 1988).

The results of Petz and Foissner (1989b) were confirmed by Wanner and Funke (1989) and Wanner (1991), who investigated the effects of lindane and Ripcord (insecticides against bark beetles) on the testate amoebae of a strongly acidic spruce forest soil in Germany (Table 7.5). These authors

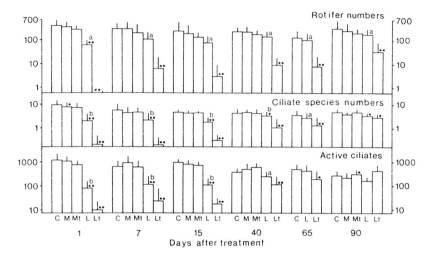

**Fig. 7.7.** Arithmetic means (± SD) of the abundance and species number of active ciliates and of numbers of rotifers at 0–33mm soil depth following pesticide treatments. Control (C); standard dose: mancozeb (M), lindane (L); high dose: mancozeb (Mt), lindane (Lt). Numbers (log e) are given as individuals g⁻¹ dry soil. *0.05 < P ≤ 0.1, **P ≤ 0.05, differences from control; a, 0.05 < P ≤ 0.1; b, P ≤ 0.05, differences from high dose. (From Petz and Foissner, 1989b.)

used, however, a simpler experimental design than Petz and Foissner and the first samples were taken only 20 months after application of the biocides.

Rather different results were reported by Laminger and Maschler (1986), who investigated the effects of some biocides (mancozeb, parathion, paraquat) on the protozoa of a subalpine arable soil in Austria. This paper, however, is difficult to understand and very probably contains some major mistakes. Laminger and Maschler (1986) themselves suggest that their control plots were contaminated by the biocides and accordingly used an earlier investigation at another site as control plots (column 'a' in their figure 2). In my experience, it is impossible to substitute controls in this manner, because abundances and dominance structures of soil protozoa usually fluctuate so distinctly that treatment effects could be confounded.

Most of the data reviewed in the following paragraphs do not consider effects on species level and have used Singh's culture method without distinguishing between active and cystic individuals. Future research will decide whether such crude methods are of any value.

Popovici *et al.* (1977) studied the effects of the common herbicide atrazine on the soil protozoa of a maize field in Romania (Table 7.11). They found a strong, dose-dependent decline of the individual numbers, especially of the flagellates. The inhibiting effect was still present four

**Table 7.9.** Percentage of ecological groups of ciliates 1 and 90 days after treatment with mancozeb and lindane at normal (0.096 g m$^{-2}$ and 6 g m$^{-2}$ active ingredient, respectively) and high (6 g m$^{-2}$ and 60 g m$^{-2}$ active ingredient, respectively) doses. (From Petz and Foissner, 1989b.)

| Ecological group | Day | Control | Mancozeb | | Lindane | |
|---|---|---|---|---|---|---|
| | | | 1 × | 10 × | 1 × | 10 × |
| Grossglocknerids | 1 | 18.5 | 9.6** | 6.7* | 15.3[+] | 0.0** |
| (mycophagous) | 90 | 14.8 | 15.2 | 20.6 | 26.0 | 39.1** |
| *Colpoda* spp. | 1 | 4.8 | 2.2 | 4.2 | 5.6 | χ[a] |
| (bacteriophagous) | 90 | 1.5 | 0.9 | 1.3 | 4.1[++] | 53.6** |
| Other colpodids | 1 | 40.2 | 55.3* | 55.3 | 26.4[++] | 0.0** |
| | 90 | 52.8 | 53.4 | 53.5** | 46.7[++] | 4.3** |
| Hypotrichs | 1 | 9.6 | 7.4**[+] | 8.7 | 6.9[+] | 0.0** |
| (omnivorous) | 90 | 19.2 | 15.2 | 11.0 | 16.6[++] | 0.8** |
| Remaining ciliates | 1 | 23.6 | 23.9** | 19.6** | 38.8 | χ[a] |
| | 90 | 10.7 | 13.9 | 10.0** | 4.1[++] | 1.0** |

[a] High value, not representative because only two individuals were found to be active.
* $0.05 < P \leqslant 0.1$; ** $P \leqslant 0.05$; differences from control.
[+] $0.05 < P \leqslant 0.1$; [++] $P \leqslant 0.05$; differences from high dose.

months after atrazine application. These results are at least partially supported by data of Pons and Pussard (1980), who tested six herbicides on 21 strains of naked amoebae in monoxenic cultures (Table 7.12). Siduron (up to 150 mg l$^{-1}$), neburon (up to 5 mg l$^{-1}$), diuron (up to 42 mg l$^{-1}$) and the synthetic phytohormone 2.4 D (up to 84 mg l$^{-1}$) were not toxic at the tested concentrations. Atrazine (40 mg l$^{-1}$), however, acted on two strains and 250 mg l$^{-1}$ dinitroorthocresol (DNOC) inhibited growth and multiplication in all species.

Tomescu (1977) investigated the effects of the insecticide heptachlor on the protozoa of a brown earth and an alluvial–colluvial forest soil (Fig. 7.8). The substance caused a distinct decline of the individual numbers in 0–10 cm and 10–20 cm soil depth. A considerable reduction was still recognizable on three occasions in the year after application, especially in 10–20 cm and in the brown earth soil. The decline of the amoebae and ciliates was less pronounced in the upper layer of the alluvial soil, possibly due to rapid leaching of the pesticide to deeper soil layers.

Odeyemi *et al.* (1988) tested 11 agrochemicals on various groups of soil microorganisms using soil microcosms and a standard plate-count technique. Biocides were applied at recommended rates and effects were evaluated three days after application. Pentachloronitrobenzene (PCNB) and agrosan completely eliminated the protozoa in the soil, whereas

**Table 7.10.** Arithmetic means (± SD) of the abundance of active (A), precystic (B), and cystic (C) testaceans, of the abundance (D) and the species number (E) of viable (active + precystic + cystic) individuals (g$^{-1}$ dry soil) and of the abundance of parasitized specimens (F) 15 and 90 days after treatment with mancozeb and lindane at normal (0.096 g m$^{-2}$ and 6 g m$^{-2}$ active ingredient, respectively) and high (0.96 g m$^{-2}$ and 60 g m$^{-2}$ active ingredient, respectively) doses. (From Petz and Foissner, 1989b.)

| Days after application | | Control | Mancozeb 1 × | Mancozeb 10 × | Lindane 1 × | Lindane 10 × |
|---|---|---|---|---|---|---|
| 15 | A | 7833 ± 5029 | 8200 ± 2854 | 6513 ± 2817* | 3817 ± 1636* | 6596 ± 611* |
| | B | 1988 ± 1828 | 2681 ± 2366 | 1638 ± 1534 | 2537 ± 2684 | 743 ± 1025 |
| | C | 1645 ± 2487 | 1720 ± 1696 | 1383 ± 597 | 1203 ± 570⁺ | 182 ± 446 |
| | D | 11467 ± 3593 | 12600 ± 3076 | 9534 ± 2300 | 7557 ± 1581** | 7522 ± 5926** |
| | E | 5.2 ± 0.8 | 6.5 ± 1.3 | 5.4 ± 0.5 | 5.2 ± 0.8⁺ | 3.7 ± 1.4* |
| | F | 6342 ± 2062 | 6193 ± 4963 | 3788 ± 2361 | 6677 ± 3860 | 7465 ± 3307 |
| 90 | A | 5404 ± 4343 | 8407 ± 4639 | 6136 ± 646 | 6223 ± 2625 | 3898 ± 2956 |
| | B | 1725 ± 2443 | 2057 ± 1508 | 1906 ± 1576 | 0 ± 0 | 545 ± 970 |
| | C | 2569 ± 2108 | 3488 ± 2782 | 1781 ± 2294 | 1419 ± 664 | 1263 ± 796 |
| | D | 9698 ± 7398 | 13952 ± 4270⁺ | 9823 ± 2425 | 7641 ± 2873 | 5706 ± 4212 |
| | E | 5.4 ± 1.8 | 6.4 ± 2.1 | 4.8 ± 1.2 | 4.5 ± 1.3⁺ | 2.8 ± 1.2** |
| | F | 3539 ± 1260 | 5160 ± 4833 | 3582 ± 2574 | 4819 ± 1730 | 5560 ± 2200 |

\* $0.05 < P \leq 0.1$; \*\* $P \leq 0.05$; differences from control.
⁺ $P \leq 0.05$; differences from high dose.

**Table 7.11.** Effects of the herbicide atrazine on protozoan populations (averages g$^{-1}$ moist soil)[a] (from Popovici *et al.*, 1977).

| Sampling | | Control | Atrazine | |
|---|---|---|---|---|
| Time after treatment | Organisms | Control | 5 kg ha$^{-1}$ | 8 kg ha$^{-1}$ |
| One month | Flagellates | 16000 | 4000 | 700 |
| | Naked amoebae | 470 | 140 | 110 |
| | Ciliates | 170 | 120 | 110 |
| Four months | Flagellates | 1800 | 920 | 260 |
| | Naked amoebae | 260 | 120 | 170 |
| | Ciliates | 1400 | 540 | 210 |

[a] The individual numbers were estimated with a modified Singh (1946) method. Four replicate plots were sampled.

**Table 7.12.** Effects of the herbicides atrazine and dinitroorthocresol (DNOC) on naked amoebae in monoxenic cultures. + growth (multiplication), – no growth (from Pons and Pussard, 1980).

| Species (Strain) | Atrazine (mg l$^{-1}$) | | DNOC (mg l$^{-1}$) | | |
|---|---|---|---|---|---|
| | 40 | 70 | 10 | 50 | 250 |
| *Gephyramoeba delicatula* | – | – | – | – | – |
| *Thecamoeba similis* (Pel) | + | + | – | – | – |
| *Vannella* sp. (Ru) | + | + | + | – | – |
| *Platyamoeba placida* | + | + | + | – | – |
| *Saccamoeba* sp. (Ir F1d) | Not tested | | + | + | – |
| *Thecamoeba granifera* (Tr75–S2) | + | + | – | – | – |
| *Hartmannella vermiformis* | + | + | – | – | – |
| *Vahlkampfia avara* | + | + | – | – | – |
| *Naegleria gruberi* 1518/1d CCAP | + | + | – | – | – |
| *Naegleria gruberi* 113/1 | – | – | – | – | – |
| *Acanthamoeba comandoni* | + | + | + | + | – |
| *Acanthamoeba castellanii* strain Neff | + | + | + | – | – |
| *Acanthamoeba mauritaniensis* | + | + | + | + | – |
| *Acanthamoeba polyphaga* | + | + | + | + | – |
| *Acanthamoeba quina* | + | + | + | + | – |
| *Acanthamoeba rhysodes* | + | + | + | – | – |
| *Acanthamoeba divionensis* | + | + | + | – | – |
| *Acanthamoeba paradivionensis* | + | + | + | – | – |
| *Acanthamoeba triangularis* | + | + | + | + | – |
| *Acanthamoeba lenticulata* | + | + | + | + | – |
| *Acanthamoeba culbertsoni* | + | + | + | – | – |

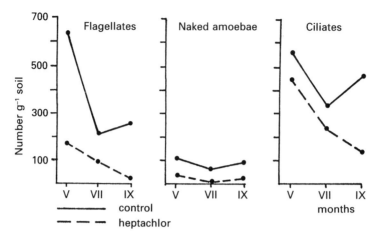

**Fig. 7.8.** Effects of the insecticide heptachlor (1 kg ha⁻¹) on the protozoa of a brown earth forest soil. Numbers were estimated with a modified Singh (1946) method in 10–20 cm soil depth three times in the year after application of the biocide and are the means from three plots (replicates). (From Tomescu, 1977.)

thiram and gammalin 20 reduced protozoa from 200 to 20 and 81 individuals g⁻¹ soil, respectively. Benlate, brestan and vetox 85 slightly depressed the protozoan population; the herbicides (gramoxone, dacthal, preforan, dual) did not have any adverse effect. This study showed that protozoa and fungi are more susceptible to pesticides, especially fungicides and insecticides (PCNB, gammalin 20, agrosan) than bacteria and actinomycetes.

Ingham and Coleman (1984) and Ingham *et al.* (1986) tested several biocides and biocide combinations on soil protozoa in microcosms under laboratory conditions (Table 7.13). Chloroform and a combination of streptomycin (bactericide) and fungizone (fungicide) produced the most distinct decline of the protozoa, possibly due to direct action (chloroform) or by reduction of the food organisms (streptomycin + fungizone). Reactions to other treatments and under field conditions (Ingham *et al.*, 1989; Fig. 7.9) varied rather distinctly, possibly due to the methodological problems inherent to Singh's culture method and the effects of ciliatostasis (Foissner, 1987a, 1989); furthermore, the litter, which contains most protozoa, was removed in these experiments.

Rather varied effects of some pesticides on the soil protozoa of a forest were also observed by Laskauskaite (1982). As this paper is written in Russian, I am unable to review it adequately.

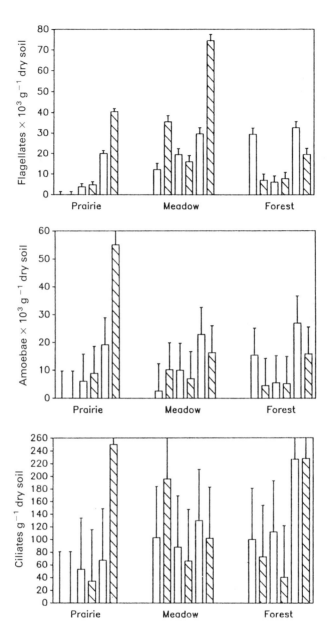

**Fig. 7.9.** Total (active + cystic; Singh's (1946) culture method) flagellate, amoebal and ciliate numbers in biocide (carbofuran and dimethoate; cross-hatched bars) and control (white bars) treatments in the upper soil layer (0–5 cm; litter removed) of a prairie, a meadow, and a pine forest on three sample dates. The first pair of bars within a group of three pairs represents the first sample values (about 6 months after biocide application), the second pair represents the second sample date values (about 8 months after biocide application), and the third pair represents the third sample date values (about 12 months after biocide application). (From Ingham et al., 1989.)

*W. Foissner*

**Table 7.13.** Effects of biocides on soil protozoa (from Ingham and Coleman, 1984, and Ingham *et al.*, 1986).

| Biocides[a] (concentration g$^{-1}$ soil) | Target group | Protozoa[b] | | |
| --- | --- | --- | --- | --- |
| | | Naked amoebae | Flagellates | Ciliates |
| Streptomycin (1 mg) | Bacteria | NE | NE | DEC (1/5) |
| Streptomycin (3 mg) | Bacteria | NE | NE | NE |
| Streptomycin (3 mg) | Bacteria | INC (2/5) | INC (1/5) | DEC (2/5) |
| Streptomycin (30 mg) | Bacteria | DEC (2/5) | DEC (4/5) | NE |
| Cycloheximide (1 mg) | Fungi | VAR | NE | VAR |
| Fungizone (1.2 mg) | Fungi | NE | NE | DEC (1/5) |
| Fungizone (1.2 mg) | Fungi | DEC (2/5) | NE | DEC (2/5) |
| Fungizone (12 mg) | Fungi | INC (2/5) | NE | DEC (2/5) |
| PCNB (quintozene; 100 μg) | Fungi | NE | DEC (1/4) | NE |
| Captan (25 μg) | Fungi | NE | NE | NE |
| Cygon (0.2 mg) | Acari | NE | NE | DEC (2/5) |
| Diazinon (47 μg) | Insects | NE | DEC (4/4) | DEC (2/4) |
| Carbofuran (25 μg) | Insects + nematodes | NE | NE | NE |
| Chloroform (1 ml) | Protozoa + bacteria | DEC (5/5) | DEC (5/5) | DEC (5/5) |
| Cygon + carbofuran + chloroform (0.2 mg, 25 μg, 1 ml) | Acari + insects + protozoa | DEC (5/5) | DEC (5/5) | DEC (5/5) |
| Streptomycin + fungizone (3 mg, 1.2 mg) | Bacteria + fungi | DEC (3/5) | DEC (5/5) | DEC (1/5) |

[a] Microcosms with 10 g air-dried soil were used and sampled on days 1, 4, 6–8, 14–15 and 22–25. Individual numbers were estimated by the culture technique of Singh (1946).
[b] DEC, decreased with respect to control ($P \leqslant 0.05$); INC, increased with respect to control ($P \leqslant 0.05$); NE, no effect of biocide, i.e. treatment not significantly different from control; VAR, significant increase at one time, decrease at another, to give variable response. Numbers in parentheses indicate number of sample dates per total number of sample dates the decrease or increase occurred.

## Effects of industrial depositions and heavy metals

Tirjaková and Matis (1987) investigated the moss ciliates in polluted and relatively unpolluted regions in Czechoslovakia. A total of 58 species was found in polluted Bratislava and 105 species in relatively unpolluted Slovenia. There was some correlation between the number of species and the level of pollution in Bratislava. An increased proportion of morphological abnormalities was noticed at the very highly polluted sites; unfortunately, Tirjaková and Matis (1987) did not describe these abnormalities in detail.

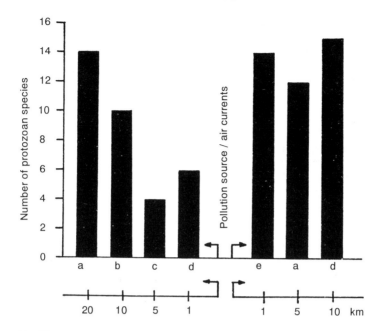

**Fig. 7.10.** Number of protozoan species in forest soils polluted by industrial emissions. Samples were taken at different distances from the pollution source and under different vegetation. a = *Quercus petraea*, b = *Fagus–Carpinus*, c = *Quercus robur*, d = *Fagus sylvatica*, e = *Pinus nigra*. (After Tomescu, 1987.)

Their species list also contains obvious misidentifications and the data are generally not presented in great detail. However, they look interesting and agree with those by Tomescu (1987), who investigated forest soil protozoa in a heavily polluted (SO$_2$, SO$_3$, Pb, Cd, Zn, Mn, Fe) region in Romania. The species number and diversity were correlated with both the distance from the source of aerial pollution and the prevailing wind direction (Fig. 7.10). In the most heavily polluted site only *r*-selected colpodids (*Colpoda inflata, Cyrtolophosis* sp., *Platyophrya vorax, Woodruffia rostrata*; see Foissner, 1993, for taxonomy and general ecology of colpodid ciliates) and some flagellates (*Cercobodo crassicauda, C. longicauda*) occurred. A total of only 25 protozoan species (flagellates, amoebae, ciliates) were found, indicating that the method used (agar and soil extract) was very inefficient. Furthermore, only few of the sites investigated had the same vegetation; differences could thus be confounded partially by differences in natural biotopes.

Even more doubtful are the results by Sztrantowicz (1980, 1984, 1987), who investigated the forest soil protozoa in an industrial region in Poland polluted by coal extractives and industrial emissions (dust, sulfur oxide,

**Table 7.14.** Infection (I) of soil invertebrates (N, number of specimens examined) with parasitic protozoa (Sporozoa) in Tanzania (7 unpolluted sites), Thailand (9 slightly polluted sites), and Germany (20 sites with high $SO_2$-deposits). (From Purrini, 1983.)

| Soil invertebrates | Tanzania | | Thailand | | Germany | |
|---|---|---|---|---|---|---|
| | N | I (%) | N | I (%) | N | I (%) |
| Enchytraeidae | 47 | 6 | 53 | 22 | 2600 | 70 |
| Lumbricidae | 72 | 7 | 66 | 12 | 1800 | 30 |
| Oribatidae | 981 | 10 | 318 | 20 | 5500 | 50 |
| Other Acarina | 199 | 2 | 88 | 12 | no data | no data |
| Collembola | 597 | 0 | 440 | 20 | 4500 | 30 |
| Other Apterygota | 193 | 0 | 89 | 17 | no data | no data |
| Myriapoda | 60 | 0 | 246 | 14 | no data | no data |
| Other Arthropoda | 175 | 8 | 135 | 16 | 2000 | 20 |

nitric oxide, heavy metals). Only 4–17 species of protozoa per site (testate and naked amoebae + flagellates + ciliates) were found, clearly indicating the incompleteness of these studies; no statistics were provided nor were active and cystic protozoa distinguished. I therefore refrain from a detailed discussion and mention only the last sentence from her 1987 paper: 'The obtained results have not revealed any helpful microbiological indicators of a degree of degradation of the environment'.

Much more convincing are the data by Purrini (1983), who made a survey on the infection of soil macro- and mesofauna with parasitic protozoa in virgin biotopes in Tanzania, in slightly polluted sites in Thailand and in heavily polluted ($SO_2$) sites in Germany. His results strongly suggest that $SO_2$-pollution dramatically increases the infection of soil invertebrates with parasitic protozoa (Table 7.14) and can be compared with the results obtained with earthworms exposed to the herbicide zeazin 50 described above (Pižl, 1985).

Pratt *et al.* (1988) evaluated the potential toxicity of leachates from several soils contaminated with organics (polycyclic aromatic hydrocarbons, e.g. fluoranthene, benzene) and/or heavy metals. Acidified tap water (pH 4.5) was used to elute toxic materials from soil columns. The leachates were used as complex mixtures in acute toxicity tests using *Daphnia* and in chronic toxicity tests using aquatic microcosms. Three classes of effects were observed. Three soil leachates showed acute and chronic toxicity at less than 3% leachate. Two of these soils were contaminated by substances used in wood preservation, and the third soil was contaminated with heavy metals and acid. Two soils showed moderately acute toxicity, but no chronic toxicity in microcosm tests. One of these soils was contaminated with low levels of chromium while the other soil was from a coal storage area (Fig.

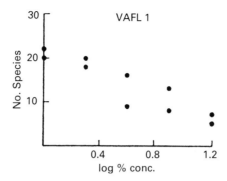

**Fig. 7.11.** Protozoan species survival in microcosms developed with leachates from hazardous and toxic waste sites. Control values with no leachate are plotted on the ordinate. VAFL1 = soil contaminated by heavy metals (e.g. Cu, Cd, Ni, Pb, Zn). (From Pratt *et al.*, 1988.)

7.11). The remaining two soil samples showed no toxicity in either acute or chronic toxicity tests. One of these soils was from an agricultural field used as a control while the other soil was contaminated with solvents. The failure to detect toxicity in the solvent-contaminated sample was attributed to the hydrophobicity of the toxic material. Results of these toxic screenings are in the same range as leachate toxicities estimated using other methods, although other methods use extraction materials that may interfere with some biological tests. Pratt *et al.* (1988) suggest that toxicological evaluations should be used in remedial studies of cleaned sites to ensure that toxic materials have been effectively removed from the site.

The approach by Pratt *et al.* (1988) is very useful and promising. The only problem is that they used aquatic rather than terrestrial protozoans as test organisms. In my experience, limnetic ciliates survive poorly in soils (see Table 15 in Foissner, 1987a), and soil protozoa might have a different sensitivity to test substances. For example, the soil ciliate *Oxytricha granulifera* Foissner is much more resistant to cadmium than the freshwater species, *Stylonychia lemnae* Ammermann and Schlegel (Irato *et al.*, 1991).

Very recently, Forge *et al.* (1991) used a similar method to Pratt *et al.* (1988) to study the toxicity of heavy metals with sewage sludge. Soil solutions obtained by centrifugation from experimentally contaminated soil were tested against the soil ciliate, *Colpoda steinii* Maupas (Table 7.15). No significant growth inhibition occurred with the February soil solutions. In July, however, when the soil solutions contained concentrations of zinc and nickel within the inhibitory ranges obtained in laboratory experiments with standard sulfate salt solutions, growth of *C. steinii* was significantly depressed. For all metals tested in sulfate solutions in the laboratory, growth of *C. steinii* was significantly reduced at 1.0 ppm ($P \leqslant 0.01$). This indicates

**Table 7.15.** Numbers of the ciliate *Colpoda steinii* after growth for 24 hours in soil solutions from experimental plots that received sewage sludge and heavy metals to result in metal concentrations at two times the maxima recommended in recent EC guidelines. (From Forge *et al.*, 1991, and unpublished.)

|  | February 1991 | | July 1991 | |
|---|---|---|---|---|
|  | Metal (mg l⁻¹) | *C. steinii* (cells ml⁻¹)[a] | Metal (mg l⁻¹) | *C. steinii* (cells ml⁻¹)[a] |
| Cadmium | 0.007 | 118500 | 0.01 | 86000 |
| Chromium | 0.209 | 100600 | 0.194 | 89400 |
| Copper | 0.115 | 85000 | 0.257 | 89200 |
| Lead | 0.291 | 105000 | 0.106 | 84500 |
| Nickel | 0.287 | 95400 | 0.859 | 60700 |
| Zinc | 1.310 | 99800 | 2.693 | 65000 |
| Control Soil | 0.0 | 90400 | 0.0 | 97500 |

[a] In February, treatments and controls do not differ ($P \geqslant 0.05$). In July, nickel and zinc treatments differ from controls ($P \leqslant 0.05$, ANOVA).

that *C. steinii* is more sensitive to heavy metals than other soil organisms that have been tested, but its sensitivity is similar to that of organisms used for monitoring pollution of freshwaters, such as *Daphnia magna*, *Gammarus pulex* and other protozoans. This protozoan bioassay thus appears to be faster and more sensitive than the existing earthworm protocols officially approved by the EC.

### Effects of road traffic

Roadsides are often polluted by heavy metal and polycyclic products from combustion engines. Grass verges along busy motorways should obviously not be used as fodder for milking cows. Only two preliminary studies are available on the protozoa of such biotopes. Lüftenegger and Foissner (1989b) studied the soil fauna in two transects at right angles to a major motorway in Austria. There was only a slight increase of heavy metals and organics (polycyclic aromatic hydrocarbons, e.g. pyrene) in the soil near the motorway and the soil fauna was richest at those sites of the transects having the highest level of total organic carbon, Pb and organics. The lowest abundances and biomasses of testate amoebae, nematodes and earthworms occurred, however, at the sites that were nearest to the motorway (Table 7.16).

Balik (1991) studied the soil testate amoebae from ten localities with different levels of road traffic pollution in Warsaw (Poland). He found a distinct loss of individuals and species near the most frequented roads (sites 1, 2, 3 in Table 7.17) as compared to natural meadows and forest soils

**Table 7.16.** Pedological and soil zoological parameters at two transects on meadows near a motorway (from Kasperowski and Frank, 1989, Lüftenegger and Foissner, 1989b).

| Parameters | Site number[a] | | | | | |
| --- | --- | --- | --- | --- | --- | --- |
| | 7a | 7 | 8 | 11 | 12 | 12a |
| Total organic carbon (%) | 5.9 | 8.2 | 4.6 | 5.8 | 8.5 | 8.3 |
| Nitrogen (%) | 10.7 | 10.0 | 11.3 | 12.4 | 10.8 | 11.0 |
| Pb (mg kg$^{-1}$ DM) | 39 | 30 | 30 | 21 | 34 | 29 |
| Cd ($\mu$g kg$^{-1}$ DM) | 350 | 473 | 581 | 598 | 661 | 575 |
| Chloride (mg kg$^{-1}$ DM) | 55 | 55 | 67 | 44 | 75 | 58 |
| Polycyclic aromatic hydrocarbons (ppb) | 181 | 188 | 160 | 118 | 198 | 151 |
| Testate amoebae[b] | | | | | | |
|   Number (g$^{-1}$ DM) | 340 | 305 | 197 | 135 | 342 | 262 |
|   Biomass ($\mu$g g$^{-1}$ DM) | 9.7 | 8.9 | 2.5 | 3.0 | 6.6 | 4.3 |
|   Number of species | 20 | 19 | 15 | 10 | 20 | 19 |
| Nematodes[b] | | | | | | |
|   Number (g$^{-1}$ DM) | 167 | 190 | 101 | 95 | 220 | 211 |
| Lumbricids[c] | | | | | | |
|   Number (m$^{-2}$) | 97 | 82 | 85 | 68 | 202 | 164 |
|   Biomass (g m$^{-2}$) | 32 | 20 | 16 | 11 | 60 | 44 |

[a] The motorway is between sites 8 and 11, which are both about 10 m away from this road; sites 7 and 12 are about 50 m, and sites 7a and 12a are about 100 m away from the motorway. About 11,000 vehicles per day pass these sites.
[b] Direct counts in soil suspensions diluted with water (Lüftenegger *et al.*, 1988). DM, dry mass of soil.
[c] Formaldehyde extraction.

(sites 5, 9, 10 in Table 7.17). Small and euryoecious species were dominant in the polluted sites.

Rauenbusch (1987) observed an exceptionally high number of testate amoebae in the soil near a motorway in Germany where the pine trees were completely destroyed by NaCl applied to the roads and unnaturally high moss carpets (up to 30 cm) developed. He also noted a loss of certain genera, e.g. *Trinema*. Wanner (1991) observed a slight decrease of the testate amoebae in a NaCl-treated plot (Table 7.5).

## Effects of polychlorinated biphenyls

Polychlorinated biphenyls (PCBs) are rather toxic compounds and their commercial use is controlled in many countries. Food contaminated with PCB can cause the Yusho-disease (Verband der Chemischen Industrie, 1982). Steinberg *et al.* (1990) studied the effects of a commercial PCB

**Table 7.17.** Influence of road traffic on testate amoebae. (From Balik, 1991.)

| Parameters[a] | Sites[b] | | | | | | | | | |
|---|---|---|---|---|---|---|---|---|---|---|
| | 1 | 2 | 3 | 4 | 5 | 6 | 7 | 8 | 9 | 10 |
| Number of species | 8 | 6 | 6 | 13 | 28 | 10 | 9 | 9 | 21 | 18 |
| Individuals cm$^{-2}$ | 6300 | 4200 | 3600 | 13800 | 55200 | 4500 | 13200 | 6600 | 40500 | 19200 |
| Diversity | 1.56 | 1.57 | 1.70 | 2.12 | 2.42 | 1.21 | 1.65 | 1.94 | 2.52 | 2.34 |
| Evenness[c] | 0.75 | 0.88 | 0.95 | 0.83 | 0.73 | 0.96 | 0.75 | 0.88 | 0.83 | 0.81 |

[a] Direct counts by the method of Lüftenegger *et al.* (1988).

[b] Sites 1, 2, 3 are near very busy roads; sites 6, 7, 8 are near busy roads; and sites 4, 5, 9, 10 are rather distant from roads.

[c] Diversity divided by ln of species number. Indicates whether the size of the diversity index is caused by an even distribution of individuals or by a high number of species. See textbooks and Wodarz *et al.* (1992) for further explanation and problems with the calculation of diversity indices.

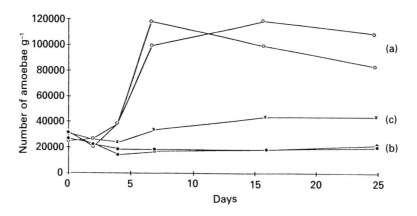

**Fig. 7.12.** Effect of pyralene on the growth of indigenous naked soil amoebae in microcosms (two replicates each). Numbers (active + cystic) were estimated with Singh's (1946) culture method. a = microcosms inoculated with *Azospirillum lipoferum*, b = microcosms inoculated with *A. lipoferum* and contaminated with 2500 ppm pyralene, c = control (i.e. not inoculated with bacteria and not contaminated with pyralene). (From Steinberg *et al.*, 1990.)

product, pyralene (Prodelec, France; PCBs in trichlorobenzene) on the predator/prey relationship of bacteria and naked soil amoebae. In the absence of pyralene, the inoculated bacterial population decreased from $10^7$ to $10^4$ cells $g^{-1}$ dry mass of soil, i.e. was grazed by the indigenous amoebae whose number increased three-fold. In the presence of 2500 ppm pyralene, the introduced bacteria (*Azospirillum lipoferum*) survived at a higher level ($3 \times 10^6$ bacteria $g^{-1}$ dry soil) while the number of amoebae decreased slightly (Fig. 7.12). The indigenous bacterial microflora was not affected quantitatively by pyralene. Bacterial growth was inhibited and amoebae were killed in pure cultures containing 2500 ppm pyralene. Steinberg *et al.* (1990) conclude from these experiments that the active amoebae could not encyst and were killed in the contaminated microcosms.

### Effects of oil pollution

The data reviewed in Foissner (1987a) indicate that crude oil does not damage the protozoan fauna of soil and that ciliates enhance the *in vitro* microbial degradation of crude petroleum; the ciliate populations in tidal sand-flats are, however, seriously disturbed by crude oil. The resistance of soil protozoa to oil stress is confirmed by a more recent study (Borisovich, 1985). The total abundance and the species richness of the protozoa were usually higher in the oil-contaminated plots than in the controls and even

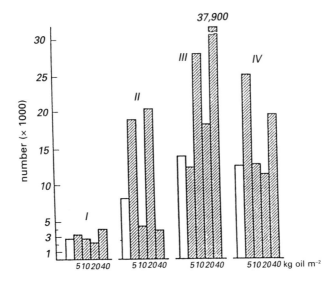

**Fig. 7.13.** Effects of various amounts of oil on the total number (g⁻¹ dry soil) of soil protozoa. Control: unhatched; oil-contaminated: hatched. I = three days after contamination, II = two months after contamination, III = 11 months after contamination, IV = three years after contamination. (From Borisovich, 1985.)

high doses of oil had no detrimental effects (Fig. 7.13). The microbial degradation of the oil could be enhanced by fertilizing the plots with farmyard manure, urea, lime and/or superphosphate.

### Effects of radioactive pollution

Most protozoa are highly resistant to ionizing radiation, viz. doses of 100–300 Ci do not produce marked effects in laboratory experiments. However, some spathidid and colpodid soil ciliates are rather sensitive, showing reduced survival at 25 kR (Foissner, 1987a). Literature on radioactive pollution and soil protozoa is sparse. The following review is mainly based on the account by Krivolutsky *et al.* (1983) who summarize a Russian experiment; see also Foissner (1987a) for further information.

300 Ci did not influence *Amoeba terrestris*, but 500 Ci produced some excitation. 1–10 Ci Co⁶⁰ γ-radiation increased the motility and division rate in the soil ciliates *Colpoda maupasi* and *C. steinii* while 50–100 Ci decreased them; 300–500 Ci was lethal for some specimens depending on the dosage. Treatment with doses of up to 100 Ci (at a rate up to 36,000 R/min) caused death of the ciliates within 3–4 days.

**Table 7.18.** Species number and total (active + cystic) abundance of protozoa in an experimental plot polluted with $Sr^{90}$ (2–3 Ci $m^{-2}$). (From Krivolutsky *et al.*, 1983.)

| Protozoa[a] | Polluted plot | | Control | |
| --- | --- | --- | --- | --- |
| | Species number | Abundance | Species number | Abundance |
| Flagellates | 8 | 17250 | 9 | 165655 |
| Naked amoebae | 8 | 7170 | 15 | 183785 |
| Ciliata | 1 | 25 | 1 | 45 |
| Total | 17 | 24445 | 25 | 349485 |

[a] Singh's (1946) culture method was used.

Korganova (1973) and Krivolutsky *et al.* (1983) studied the protozoa in the upper layer (0–5 cm) of a chernozem meadow soil experimentally polluted with $Sr^{90}$. In spite of the above-mentioned resistance there was a rather distinct decline in the number of individuals and species in the polluted site possibly due to the enrichment of $Sr^{90}$ in the water film surrounding the contaminated soil particles (Table 7.18).

## Recolonization of heavily disturbed soils

Re-establishment of biological activity in soils that have been subjected to gross physical or chemical disturbance is desirable as soon as possible for productivity and for aesthetic reasons. The use of indicator organisms or biochemical indices have been suggested as rapid techniques for assessing the recovery of 'soil health' following a disturbance. With regards to animals, most studies have been concerned with the meso- and macrofauna. There are, however, a few protozoological studies. Most were initiated in response to Russell and Hutchinsons' hypothesis that 'sick' infertile soils were caused by protozoan grazing on bacteria. These studies were reviewed by Foissner (1987a) and showed that recolonization of partially sterilized soils (by formalin or steam) occurs in a few months. This is substantiated by a recent, well-designed study that assessed the course of recovery in biological activity in the top 5 cm of soil cores (30 cm diameter, 30 cm deep) that had been fumigated in the laboratory with methyl bromide (Bamforth and Yeates, 1991; Yeates *et al.*, 1991). The cores were returned to their original pasture and forest sites, and untreated soils at all sites served as controls. Sampling took place 0, 1, 5, 12, 26, 54, 110 and 126 days after fumigation. Fumigation almost totally eliminated protozoa; only the flagellates *Oikomonas termo* and *Pleuromonas jaculans* and a small vahlkampfid amoeba

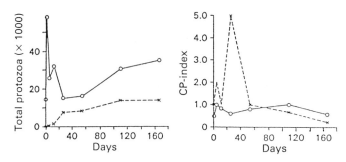

**Fig. 7.14.** Changes in the individual abundances (g⁻¹ dry mass of soil) of protozoa and the CP-index (ratio of colpodid/polyhymenophoran ciliates) in an untreated (o——o) and a fumigated (x----x) pasture soil over 166 days. Each point is the mean of three replicates. Active + cystic naked amoebae, flagellates, and ciliates were estimated by the culture technique of Singh (1946); testate amoebae were counted in soil suspensions diluted with water. (From Yeates *et al.*, 1991.)

survived. *Bodo globosus, B. mutabilis, Colpoda inflata, C. steinii, Cryptodifflugia compressa, Difflugiella oviformis, Microcorycia flava, Platyamoeba placida, Tracheleuglypha acolla* and *Trinema lineare* were the earliest colonists. The proportion of earliest colonists to total species decreased from 100% to 50% by day 26; and the colpodid/polyhymenophoran ratio (CP-index; see Lüftenegger *et al.*, 1985) decreased below 1, suggesting a return to the original conditions (Fig. 7.14). Species richness was restored to the original by day 110.

A similar experiment was performed by Palka (1991) using, however, laboratory soil microcosms and three methods of soil sterilization (propylene oxide, autoclaving, γ-irradiation). According to this author, three groups of ciliates can be distinguished: indifferent species (*Colpoda aspera, C. inflata, C. steinii, Gonostomum affine*), semisensitive species (*Chilodonella uncinata, Holosticha adami, H. multistilata, Spathidium* sp.), and sensitive species (*Blepharisma* sp., *Drepanomonas* sp., *Urosomoida agiliformis*). These results support the suggestion of Lüftenegger *et al.* (1985) and Wodarz *et al.* (1992) that colpodids are more *r*-selected than spirotrichs.

Recolonization is slow after topsoil removal. Five to seven years after smoothing mountain ski slopes, the abundance and biomass of testaceans and nematodes were still approximately ten times lower in the centre of the ski slopes than in the neighbouring alpine pasture (for review see Foissner, 1987a). Recolonization with ciliates and nematodes can be significantly ($P \leqslant 0.05$) stimulated by organic fertilizers (Lüftenegger *et al.*, 1986). The ciliate numbers in the organically fertilized plots did not differ significantly ($P \geqslant 0.1$), but the Weighted Coenotic Index (WCI) shows that dried fungal biomass is more effective than dried bacterial biomass (Table 7.19). The

**Table 7.19.** Testate amoebae (upper rows) and ciliates (lower rows) in the soils of levelled, revegetated and fertilized mountain ski slopes. (From Lüftenegger *et al.*, 1986.)

| Treatments[a] | *N*[b] | *S*[c] | WCI[d] | H[e] |
|---|---|---|---|---|
| Undisturbed alpine mat | 540 | 23 | 87 | 2.628 |
| | 2 | 5 | 874631 | 1.495 |
| Dried fungal biomass | 9 | 2 | 440000000 | 0.693 |
| | 332 | 18 | 83 | 1.988 |
| Dried bacterial biomass | 1 | 1 | calculation | useless |
| | 225 | 17 | 237 | 2.153 |
| NPK fertilizer | 1 | 1 | calculation | useless |
| | 57 | 21 | 1614 | 2.525 |
| Revegetated, unfertilized | 0 | 0 | calculation | useless |
| | 5 | 9 | 425863 | 1.642 |

[a] Samples were collected three years after levelling, revegetation, and fertilization.
[b] *N*, individuals g⁻¹ dry mass of soil; direct counts with the method of Lüftenegger *et al.* (1988). Values are arithmetic means from five sampling occasions in September 1985.
[c] Total number of species.
[d] WCI, Weighted Coenotic Index (Wodarz *et al.*, 1992). With testate amoebae, decreasing WCI values indicate improving soil conditions. With ciliates, which are inhibited by the effects of ciliatostasis in natural, evolved soils (Foissner, 1987a), decreasing WCI values usually indicate nullification of ciliatostasis and pioneer soils, respectively.
[e] Shannon–Wiener's diversity index.

marked difference between the organic and mineral fertilizers is well expressed by the Weighted Coenotic Index, whereas the Shannon-Wiener index is highest in the site fertilized with NPK. Although ciliate abundance is low, this is due to the unnatural, almost even distribution of most species (high evenness). The high value of the WCI in the revegetated, unfertilized plot is obviously caused by shortage of food. The community structure of the ciliates and the low abundance of the testaceans indicate that despite three years of recultivation the soil fauna in the levelled ski slopes was still far from natural, although there was a trend in this direction, particularly in the organically fertilized plots (Lüftenegger *et al.*, 1986).

Rosa (1956) investigated the micro-edaphon in the soil cover of slowly burning brown coal dumps. He found many species of protozoa and noted that algae colonized depleted areas as soon as the temperature had fallen below 52°C.

Overall, the results by Lüftenegger *et al.* (1986) and Yeates *et al.* (1991) suggest that protozoan and nematode populations could provide a useful medium-term ecological index of the recovery in comprehensive soil biological activity following major soil pollution or disturbance.

# Summary and Conclusions

Protozoa have several unique characteristics favouring their use as bioindicators in natural ecosystems and those disturbed by man, viz., rapid growth, delicate external membranes, eukaryotic genomes, large numbers even in such ecosystems that are almost or completely devoid of higher organisms due to extreme environmental conditions (e.g. polar regions, deserts) and an almost stable and ubiquitous distribution.

This review is concerned mainly with literature after 1985 dealing with soil protozoa as bioindicators in ecosystems under human influence and the associated methodological problems. The data available show that protozoa rapidly reflect changes in their environment by changes in their population density, number of species and dominance structure.

Soil cultivation, organic farming and fertilizers usually increase the number of soil protozoa as compared with virgin lands and fallows. Prudent applications of slow-acting pH-regulators (e.g. magnesite) and natural (e.g. farmyard manure) and synthetic (e.g. dried bacterial biomass) fertilizers do not appear to disturb the soil fauna and can be recommended when the vitality and yield of crops are increased and there is no ensuing groundwater pollution.

Current evidence suggests that soil protozoa are at least as sensitive to environmental hazards (pesticides, heavy metals, etc.) as more commonly used test organisms (e.g. earthworms). There is thus a strong likelihood that protozoa can replace vertebrates in some assays. Likewise, protozoa are very rapid indicators of the recovery of biological activity in soils that have been subjected to gross physical or chemical disturbance.

Methodological problems still delay progress in soil protozoology and interfere with their use as bioindicators. The methods available for estimating the numbers of *active* soil protozoa are either rather time-consuming (direct counting in diluted suspensions) and/or unreliable (e.g. dilution culture methods). Direct examination of hundreds of fresh soil samples from various regions and under different moisture conditions showed that most protozoa are inactive (cystic) in evolved soils, except testate amoebae, which are probably the most important soil protozoans (Foissner, 1987a). Methodological problems are, however, not confined to soil protozoa (as widely assumed), but occur also in more familiar groups of soil animals, e.g. earthworms (Ehrmann and Babel, 1991).

Most studies that have used protozoa for assessing side-effects of human activities have been restricted to population densities, although there is evidence that on many occasions the community structure and species present are affected. The current inadequacy of soil protozoan systematics and lack of modern identification keys of soil protozoa seriously handicap further use of protozoa as bioindicators.

Overcoming the methodological problems is a critical requirement for such problems as teratologic testing (for a review see Schardein, 1988) and adaptation to stress (for reviews see Cairns and Niederlehner, 1989, and Nilsson, 1989). Well-designed field experiments are still rare and should help to make soil protozoology more reputable. There is, in my opinion, a genuine readiness by ecologists, governments and private companies to use protozoa in the future as rapid indicators for the re-establishment of biological activity in heavily polluted or disturbed soils and in assays for hazardous materials like pesticides and heavy metals.

# Acknowledgements

This review was supported by the Fonds zur Förderung der wissenschaftlichen Forschung (Projekt P 5889) and by the Biochemie Kundl/Tirol. I gratefully acknowledge Mag. E. Strobl's and Dr A. Unterweger's help in improving my English and in typing the manuscript, respectively. My special thanks to the volume editor, Dr J.F. Darbyshire, for his advice on the content of the manuscript and English syntax.

# References

Aescht, E. and Foissner, W. (1991) Bioindikation mit mikroskopisch kleinen Bodentieren. *VDI Berichte* 901, 985–1002.
Aescht, E. and Foissner, W. (1992) Effects of mineral and organic fertilizers on the microfauna in a high altitude afforestation trial. *Biology and Fertility of Soils* 13, 17–24.
Aescht, E. and Foissner, W. (1993) Effects of organically enriched magnesite fertilizers on the soil ciliates of a spruce forest. *Pedobiologia* 37, 321–335.
Balik, V. (1991) Die Strassenverkehrseinwirkung auf die im Boden lebenden Testaceenzönosen (Rhizopoda, Testacea) in Warschau (Polen). *Acta Protozoologica* 30, 5–11.
Bamforth, S.S. and Yeates, G.W. (1991) Aerial colonization by protozoa of methyl bromide sterilized soils. *Journal of Protozoology* 38, Supplement 12A (abstract 71).
Beck, L., Dumpert, K., Franke, U., Römbke, J., Mittmann, H.-W. and Schönborn, W. (1988) Vergleichende ökologische Untersuchungen in einem Buchenwald nach Einwirkung von Umweltchemikalien. *Spezielle Berichte der Kernforschungsanlage Jülich* 439, 548–701.
Berger, H., Foissner, W. and Adam, H. (1985) Protozoologische Untersuchungen an Almböden im Gasteiner Tal (Zentralalpen, Österreich). IV. Experimentelle Studien zur Wirkung der Bodenverdichtung auf die Struktur der Testaceen- und Ciliatentaxozönose. *Veröffentlichungen des Österreichischen MaB-Programms* 9, 97–112.
Berthold, A. and Foissner, W. (1993) Singh's dilution culture method is inappropriate

for estimating individual numbers of active soil ciliates (Protozoa). *Journal of Protozoology* 40 Supplement 17A, (abstract 98).

Bick, H. (1982) Bioindikatoren und Umweltschutz. *Decheniana-Beihefte (Bonn)* 26, 2–5.

Borisovich T.M. (1985) The effect of oil pollution of soil protozoa. In: Striganova, B.R. (ed.), *Soil Fauna and Soil Fertility (Proceedings of the 9th International Colloquium on Soil Zoology)*. Nauka, Moscow, pp. 282–284 (in Russian).

Cairns, J. and Niederlehner, B.R. (1989) Adaptation and resistance of ecosystems to stress: a major knowledge gap in understanding anthropogenic perturbations. *Speculations in Science and Technology* 12, 23–30.

Cairns, J. and Pratt, J.R. (1990) Biotic impoverishment: effects of anthropogenic stress. In: Woodwell, G.M. (ed.), *The Earth in Transition: Patterns and Processes of Biotic Impoverishment.* Cambridge University Press, Cambridge, pp. 495–505.

Darbyshire, J.F., Griffiths, B.S., Davidson, M.S. and McHardy, W.J. (1989) Ciliate distribution amongst soil aggregates. *Revue d'Écologie et de Biologie du Sol* 26, 47–56.

Dive, D., Leclerc, H. and Persoone, G. (1980) Pesticide toxicity on the ciliate protozoan *Colpidium campylum*: possible consequences of the effect of pesticides in the aquatic environment. *Ecotoxicology and Environmental Safety* 4, 129–133.

Domsch, K.H., Jagnow, G. and Anderson, T.-H. (1983) An ecological concept for the assessment of side-effects of agrochemicals on soil microorganisms. *Residue Reviews* 86, 65–105.

Ehrmann, O. and Babel, U. (1991) Quantitative Regenwurmerfassung – ein Methodenvergleich. *Mitteilungen der Deutschen Bodenkundlichen Gesellschaft* 66, 475–478,

Foissner, W. (1987a) Soil protozoa: fundamental problems, ecological significance, adaptations in ciliates and testaceans, bioindicators, and guide to the literature. *Progress in Protistology* 2, 69–212.

Foissner, W. (1987b) Ökologische Bedeutung und bioindikatives Potential der Bodenprotozoen. *Verhandlungen der Gesellschaft für Ökologie (Gießen 1986)* 16, 45–52.

Foissner, W. (1987c) The micro-edaphon in ecofarmed and conventionally farmed dryland cornfields near Vienna (Austria). *Biology and Fertility of Soils* 3, 45–49.

Foissner, W. (1989) Ciliatostasis: Ein neuer Ansatz zum Verständnis terricoler Protozoen-Zönosen. *Verhandlungen der Gesellschaft für Ökologie (Göttingen 1987)* 17, 371–377.

Foissner, W. (1991) Diversity and ecology of soil flagellates. In: Patterson, D.J. and Larsen, J. (eds), *The Biology of Free-living Heterotrophic Flagellates.* Clarendon Press, Oxford, pp. 93–112.

Foissner, W. (1992) Comparative studies on the soil life in ecofarmed and conventionally farmed fields and grasslands of Austria. *Agriculture, Ecosystems and Environment* 40, 207–218.

Foissner, W. (1993) *Colpodea (Ciliophora)*. Gustav Fischer Verlag, Stuttgart, Jena, New York.

Foissner, W., Franz, H. and Adam, H. (1986) Untersuchungen über das Bodenleben in ökologisch und konventionell bewirtschafteten Acker-und Grünlandböden im Raum Salzburg. *Verhandlungen der Gesellschaft für Ökologie (Giessen 1985)* 15, 333–339.

Foissner, W., Buchgraber, K. and Berger, H. (1990) Bodenfauna, Vegetation und Ertrag bei ökologisch und konventionell bewirtschaftetem Grünland: Eine Feldstudie mit randomisierten Blöcken. *Mitteilungen der Österreichischen Boden-kundlichen Gesellschaft* 41, 5–33.

Forge, T.A., Berrow, M.L., Darbyshire, J.F. and Warren, A. (1991) Rapid protozoan assays of soil contaminated with heavy metals. *8th International Conference on Heavy Metals in the Environment, Vol. I.* CEP Consultants, Edinburgh, pp. 310–313.

Foster, R.C. and Dormaar, J.F. (1991) Bacteria-grazing amoebae in situ in the rhizosphere. *Biology and Fertility of Soils* 11, 83–87.

Funke, W. (1987) Wirbellose Tiere als Bioindikatoren in Wäldern. *VDI Berichte* 609, 133–176.

Funke, W. and Roth-Holzapfel, M. (1991) Animal-coenoses in the 'spruce forest' ecosystem (Protozoa, Metazoa–invertebrates): indicators of alterations in forest-ecosystems. In: Esser, G. and Overdieck, D. (eds), *Modern Ecology, Basic and Applied Aspects.* Elsevier, Amsterdam, pp. 579–600.

Glasbey, C.A., Horgan, G.W. and Darbyshire, J.F. (1991) Image analysis and three-dimensional modelling of pores in soil aggregates. *Journal of Soil Science* 42, 479–486.

Groliere, C.A., Sefrioui, S. and Dupy-Blanc, J. (1992) Bioconversion of thiram (TMTD) by *Tetrahymena pyriformis. Archiv für Protistenkunde* 141, 135–140.

Gupta, V.V.S.R. and Germida, J.J. (1988) Populations of predatory protozoa in field soils after 5 years of elemental S fertilizer application. *Soil Biology and Biochemistry* 20, 787–791.

Hill, S.B. and Kevan, D.K. McE. (1985) The soil fauna and agriculture: past findings and future priorities. In: Striganova, B.R. (ed.), *Soil Fauna and Soil Fertility (Proceedings of the 9th International Colloquium on Soil Zoology).* Nauka, Moscow, pp. 49–56.

Ingham, E.R. and Coleman, D.C. (1984) Effects of streptomycin, cycloheximide, fungizone, captan, carbofuran, cygon, and PCNB on soil microorganisms. *Microbial Ecology* 10, 345–358.

Ingham, E.R., Cambardella, C. and Coleman, D.C. (1986) Manipulation of bacteria, fungi and protozoa by biocides in lodgepole pine forest soil microcosms: effects on organism interactions and nitrogen mineralization. *Canadian Journal of Soil Science* 66, 261–272.

Ingham, E.R., Coleman, D.C. and Moore, J.C. (1989) An analysis of food-web structure and function in a shortgrass prairie, a mountain meadow, and a lodgepole pine forest. *Biology and Fertility of Soils* 8, 29–37.

Irato, P., Ammermann, D. and Piccinni, E. (1991) Effects of Cd on *Oxytricha granulifera* and *Stylonychia lemnae* (Ciliophora, Hypotrichida). *Journal of Protozoology* 38, Supplement 33A (abstract 194).

Kasperowski, E. and Frank, E. (1989) Boden- und Vegetationsuntersuchungen im Bereich der Scheitelstrecke der Tauernautobahn. *Umweltbundesamt Wien, Monographie* 15, I-IV + 1–126.

Klein, M.G. (1988) Pest management of soil-inhabiting insects with microorganisms. *Agriculture, Ecosystems and Environment* 24, 337–349.

Köhler, W., Schachtel, G. and Voleske, P. (1984) *Biometrie. Einführung in die Statistik für Biologen und Agrarwissenschaftler.* Springer, Berlin, Heidelberg, New York and Tokyo.

Komala, Z. (1978) The effect of lindane on *Paramecium primaurelia. Folia Biologica (Kraków)* 26, 355–360.

Korganova, G.A. (1973) Influence of experimental soil pollution with $^{90}$Sr on the fauna of soil protists. *Zoologicheskij Zhurnal* 52, 939–941 (in Russian with English summary).

Krivolutsky, D.A., Korganova, G.A. and Tikhomirov, F.A. (1983) Some radioecolog-
ical problems of terrestrial and soil animals. In: Lebrun, P., André, H.M., De
Medts, A., Grégoire-Wibo, C. and Wauthy, G. (eds), *New Trends in Soil Biology
(Proceedings of the 8th International Colloquium of Soil Zoology)*. Dieu-Brichart, Ottignies-Louvain-la-Neuve, pp. 463–469.

Kuikman, P.J., Van Elsas, J.D. Jansen, A.G., Burgers, S.L.G.E. and Van Veen, J.A.
(1990) Population dynamics and activity of bacteria and protozoa in relation
to their spatial distribution in soil. *Soil Biology and Biochemistry* 22, 1063–1073.

Kuikman, P.J., Jansen, A.G. and Van Veen, J.A. (1991) $^{15}$N-nitrogen mineralization
from bacteria by protozoan grazing at different soil moisture regimes. *Soil
Biology and Biochemistry* 23, 193–200.

Laminger, H. and Maschler, O. (1986) Auswirkungen von einigen Bioziden auf die
Bodenprotozoen im Raum Vill (Tirol/Österreich). *Zoologischer Anzeiger* 216,
109–122.

Laskauskaite, D. (1982) The effect of pesticides on the biocoenosis structure of soil
invertebrates. Protozoa. In: Eitminaviviute, I.S. (ed.), *The Effect of Pesticides
on Pedobionts and Biological Activity of Soil*. Mokslas, Vilnius, pp. 50–58 (In
Russian).

Lehle, E. and Funke, W. (1989) Zur Mikrofauna von Waldböden. II. Ciliata
(Protozoa: Ciliophora) Auswirkungen anthropogener Einflüsse. *Verhandlungen
der Gesellschaft für Ökologie (Göttingen 1987)* 17, 385–390.

Lousier, J. D. (1974a) Effects of experimental soil moisture fluctuations on turnover
rates of testacea. *Soil Biology and Biochemistry* 6, 19–26.

Lousier, J.D. (1974b) Response of soil testacea to soil moisture fluctuations. *Soil
Biology and Biochemistry* 6, 235–239.

Lousier, J.D. and Bamforth, S.S. (1990) Soil protozoa. In: Dindal, D.L. (ed.), *Soil
Biology Guide*. Wiley, New York, pp. 97–136.

Lüftenegger, G. and Foissner, W. (1989a) Bodenzoologische Untersuchungen an
ökologisch und konventionell bewirtschafteten Weinbergen. *Landwirtschaftli-
che Forschung* 42, 105–113.

Lüftenegger, G. and Foissner, W. (1989b) Bodenzoologische Untersuchungen. In:
Kasperowski, E. and Frank, E. (eds) Boden- und Vegetationsuntersuchungen
im Bereich der Scheitelstrecke der Tauernautobahn. *Umweltbundesamt Wien,
Monographie* 15, 88–93.

Lüftenegger, G., Foissner, W. and Adam, H. (1985) *r*– and *K*-selection in soil ciliates:
a field and experimental approach. *Oecologia (Berlin)* 66, 574–579.

Lüftenegger, G., Foissner, W. and Adam, H. (1986) Der Einfluß organischer und
mineralischer Dünger auf die Bodenfauna einer planierten, begrünten Schipiste
oberhalb der Waldgrenze. *Zeitschrift für Vegetationstechnik* 9, 149–153.

Lüftenegger, G., Petz, W., Foissner, W. and Adam, H. (1988) The efficiency of a direct
counting method in estimating the numbers of microscopic soil organisms.
*Pedobiologia* 31, 95–101.

Magurran, A.E. (1988) *Ecological Diversity and its Measurement*. Princeton University Press, Princeton, New Jersey.

Mathur, R. and Saxena, D.M. (1986) Inhibition of macromolecule syntheses in a
ciliate protozoan, *Tetrahymena pyriformis*, by hexachlorocyclohexane (HCH)
isomers. *Acta Protozoologica* 25, 397–409.

Mathur, R. and Saxena, D.M. (1988) Effect of hexachlorocylohexane (HCH) isomers
on cell shape and size of a freshwater ciliate, *Tetrahymena pyriformis. Journal
of Advanced Zoology* 9, 76–82.

Mordkovich, G.D. (1986) The changes in soil protozoan populations under the agricultural development of virgin lands. *Izvestiya Sibirskogo Otdeleniya Akademii Nauk SSSR (Seriya Biologicheskih Nauk)* 18, 44–49.

Nilsson, J.R. (1989) *Tetrahymena* in cytotoxicology: with special reference to effects of heavy metals and selected drugs. *European Journal of Protistology* 25, 2–25.

Odeyemi, O., Salami, A. and Ugoji, E.O. (1988) Effect of common pesticides used in Nigeria on soil microbial population. *Indian Journal of Agricultural Sciences* 58, 624–628.

Old, K.M. and Chakraborty, S. (1986) Mycophagous soil amoebae: Their biology and significance in the ecology of soil-borne plant pathogens. *Progress in Protistology* 1, 163–194.

Palka, L. (1991) Étude expérimentale de la recolonisation par les protozoaires d'un sol stérilisé. I. – Diversification du peuplement de ciliés. *Revue d'Écologie et de Biologie du Sol* 28, 125–131.

Paustian, K., Andrén, O., Clarholm, M., Hansson, A.-C., Johansson, G., Lagerlöf, J., Lindberg, T., Pettersson, R. and Sohlenius, B. (1990) Carbon and nitrogen budgets of four agro-ecosystems with annual and perennial crops, with and without N fertilization. *Journal of Applied Ecology* 27, 60–84.

Perey, J.F. (1925) Influence du milieu de culture sur les numérations de protozoaires du sol. *Compte Rendus de l'Académie des Sciences, Paris* 180, 315–317.

Petz, W. and Foissner, W. (1989a) Effects of irrigation on the protozoan fauna of a spruce forest. *Verhandlungen der Gesellschaft für Ökologie (Göttingen 1987)* 17, 397–399.

Petz, W. and Foissner, W. (1989b) The effects of mancozeb and lindane on the soil microfauna of a spruce forest: a field study using a completely randomized block design. *Biology and Fertility of Soils* 7, 225–231.

Pižl, V. (1985) The effect of the herbicide zeazin 50 on the earthworm infection by monocystid gregarines. *Pedobiologia* 28, 399–402.

Pons, R. and Pussard, M. (1980) Action des herbicides sur les amibes libres (Rhizopoda, Protozoa). Étude préliminaire. *Acta Oecologia* 1, 15–20.

Popovici, I., Stan, G., Stefan, V., Tomescu, R., Dumea, A., Tarta, A. and Dan, F. (1977) The influence of atrazine on soil fauna. *Pedobiologia* 17, 209–215.

Pratt, J.R., McCormick, P.V., Pontasch, K.W. and Cairns, J. (1988) Evaluating soluble toxicants in contaminated soils. *Water, Air, and Soil Pollution* 37, 293–307.

Purrini, K. (1983) Studies on pathology and diseases of soil invertebrates of palm- and forest soils in Tanzania and Thailand. In: Lebrun, P., André, H. M., De Medts, A., Grégoire-Wibo, C. and Wauthy, G. (eds), *New Trends in Soil Biology (Proceedings of the 8th International Colloquium of Soil Zoology)*. Dieu-Brichart, Ottignies-Louvain-la-Neuve, pp. 8–87.

Raederstorff, D. and Rohmer, M. (1987a) The action of the systemic fungicides tridemorph and fenpropimorph on sterol biosynthesis by the soil amoeba *Acanthamoeba polyphaga*. *European Journal of Biochemistry* 164, 421–426.

Raederstorff, D. and Rohmer, M. (1987b) Sterol biosynthesis via cycloartenol and other biochemical features related to photosynthetic phyla in the amoebae *Naegleria lovaniensis* and *Naegleria gruberi*. *European Journal of Biochemistry* 164, 427–434.

Rauenbusch, K. (1987) Biologie und Feinstruktur (REM-Untersuchungen) terrestrischer Testaceen in Waldböden (Rhizopoda, Protozoa). *Archiv für Protistenkunde* 134, 191–294.

Rosa, K. (1956) Mikroflora und Mikrofauna schwelender Halden bei Sokolov. *Biológica* 11, 541–546 (in Czech with German summary).

Rosa, K. (1974) Dynamics of the micro-edaphon of fertilized and unfertilized forest soils. *Sbornik Vědeckého Lesnického Ústavu Vysoké Školy Zemědělské v Praze* 17, 271–308 (in Czech with Russian, English and Germany summary).

Schardein, J.L. (1988) Teratologic testing: status and issues after two decades of evolution. *Reviews of Environmental Contamination and Toxicology* 102, 1–77.

Schnürer, J., Clarholm, M. and Rosswall, T. (1986) Fungi, bacteria and protozoa in soil from four arable cropping systems. *Biology and Fertility of Soils* 2, 119–126.

Schreiber, B. and Brink, N. (1989) Pesticide toxicity using protozoans as test organisms. *Biology and Fertility of Soils* 7, 289–296.

Schwerdtfeger, F. (1975) *Ökologie der Tiere III: Synökologie. Struktur, Funktion und Produktivität mehrartiger Tiergemeinschaften. Mit einem Anhang: Mensch und Tiergemeinschaft.* Parey, Hamburg and Berlin.

Singh, B.N. (1946) A method of estimating the numbers of soil protozoa, especially amoebae, based on their differential feeding on bacteria. *Annals of Applied Biology* 33, 112–119.

Smith, H.G. (1973) The Signy Island terrestrial reference sites: III. Population ecology of *Corythion dubium* (Rhizopoda: Testacida) in site 1. *British Antarctic Survey Bulletin* 33 and 34, 123–135.

Smith, H.G. (1985) Ecology of protozoa in Antarctic fellfields. In: Striganova, B.R. (ed.), *Soil Fauna and Soil Fertility (Proceedings of the 9th International Colloquium on Soil Zoology).* Nauka, Moscow, pp. 480–483.

Sohlenius, B. (1990) Influence of cropping system and nitrogen input on soil fauna and microorganisms in a Swedish arable soil. *Biology and Fertility of Soils* 9, 168–173.

Southwood, T.R.E. (1978) *Ecological Methods. With Particular Reference to the Study of Insect Populations*, 2nd edn. Chapman and Hall, London and New York.

Steinberg, C., Grosjean, M.C., Bossand, B. and Faurie, G. (1990) Influence of PCBs on the predator–prey relation between bacteria and protozoa in soil. *FEMS Microbiology Ecology* 73, 139–148.

Szabó, A. (1986) Correlations between the water supply and the microfauna of various types of soil. *Symposia Biologica Hungarica* 33, 269–274.

Sztrantowicz, H. (1980) Structure and numbers of soil protozoa on industrial areas of Silesia. *Polish Ecological Studies* 6, 607–624.

Sztrantowicz, H. (1984) Soil protozoa as indicators of environment degradation by industry. *Polish Ecological Studies* 10, 67–91.

Sztrantowicz, H. (1987) Occurrence of microorganisms in organic matter experimentally introduced into the soil of forest ecosystems situated in the industrialized region. *Polish Ecological Studies* 13, 29–51.

Tirjaková, E. (1988) Structures and dynamics of communities of ciliated protozoa (Ciliophora) in field communities. I. Species composition, group dominance, communities. *Biológia (Bratislava)* 43, 497–503.

Tirjaková, E. (1991) Species composition of soil communities of ciliated protozoa (Ciliophora) in relation to fertilizer type and concentration. *Acta Facultatis Rerum Naturalium Universitatis Comenianae Zoologia* 34, 39–43.

Tirjaková, E. and Matis, D. (1987) Ciliates of dry mosses in Bratislava in relation to air pollution. *Acta Facultatis Rerum Naturalium Universitatis Comenianae Zoologia* 29, 17–31.

Tomescu, R. (1977) Influenta tratamentelor cu heptaclor asupra protozoarelor din sol. *Studia Universitatis Babes-Bolyai-Biologia* 1, 53–56.

Tomescu, R. (1987) Influenta poluarii asupra protozoarelor din zona industriala Zlatna. *Studii si Cercetari de Biologie (Seria Biologie Animale) Romania* 39, 167–170.

Verband der Chemischen Industrie (1982) *Umwelt und Chemie von A-Z*, 4th edn. Herder, Freiburg.

Viswanath, G.K. and Pillai, S.C. (1968) Occurrence and activity of protozoa in soil. *Journal of Scientific Industrial Research* 27, 187–195.

Wanner, M. (1991) Zur Ökologie von Thekamöben (Protozoa: Rhizopoda) in süddeutschen Wäldern. *Archiv für Protistenkunde* 140, 237–288.

Wanner, M. and Funke, W. (1989) Zur Mikrofauna von Waldböden: I. Testacea (Protozoa: Rhizopoda) Auswirkungen anthropogener Einflüsse. *Verhandlungen der Gesellschaft für Ökologie (Göttingen 1987)* 17, 379–384.

Washington, H.G. (1984) Diversity, biotic and similarity indices. A review with particular relevance to aquatic ecosystems. *Water Research* 18, 653–694.

Weigmann, G. (1987) Fragen der Auswertung und Bewertung faunistischer Artenlisten. In: Becker, H. (ed.), *Untersuchung und Bewertung von Belastungen in Ökosystemen*. Parey, Berlin and Hamburg, pp. 23–33.

Wiger, R. (1985) Variability of lindane toxicity in *Tetrahymena pyriformis* with special reference to liposomal lindane and the surfactant tween 80. *Bulletin of Environmental Contamination and Toxicology* 35, 452–459.

Wodarz, D., Aescht, E. and Foissner, W. (1992) A Weighted Coenotic Index (WCI): description and application to soil animal communities. *Biology and Fertility of Soils* 14, 5–13.

Yeates, G.W., Bamforth, S.S., Ross, D.J., Tate, K.R. and Sparling, G.P. (1991) Recolonization of methyl bromide sterilized soils under four different field conditions. *Biology and Fertility of Soils* 11, 181–189.

# Future Prospects

J.F. Darbyshire
*Viewfield, Midmar, Inverurie, Aberdeenshire AB51 7QA, UK.*

Advances in soil protozoan ecology are hindered by our inability to allocate species names to all the protozoa present in soil, the protracted nature and lack of precision of existing counting methods, as well as ignorance of the soil environment at the microscopical scale. Several authors in this book have already remarked on these weaknesses.

Detailed ecological studies of soil protozoa are dependent on modern taxonomy (Corliss, 1992). In many instances, the original descriptions of existing species were made before the advent of the electron microscope. There is an obvious need for far more studies of soil protozoan taxonomy worldwide, using all the sophisticated techniques of light microscopy, electron microscopy and genetics that are necessary. Apart from improving understanding of protozoan phylogenetic relationships, such studies should increase our knowledge about protozoan geographical distribution in soils.

More rapid estimations of soil protozoan populations would greatly promote the study of protozoan ecology of mixed populations. Flow cytometry is a technique that simultaneously measures several features of individual cells at very high speeds (e.g. at 1000 cells per second). Correlations between these cellular features can also be calculated and displayed rapidly. Soil protozoologists have been slow to employ this technique in comparison with haematologists, immunologists or even other protozoologists (Shapiro, 1990; Burkill, 1992; Edwards *et al.*, 1992). Flow cytometers are often used with fluorescent markers of cellular activity or structure. The range of fluorochromes available is now extensive. Recently, fluorescently labelled oligonucleotide probes complementary to 16S rRNA sequences have been used with flow cytometers to analyse mixed bacterial populations of actively growing cells (Amman *et al.*, 1990), and similar studies could be made with soil protozoa.

An early use of flow cytometers to distinguish and sort mixed populations of mammalian leucocytes on the basis of a fluorescent product of intracellular peptidase activity was described by Dolbeare and Smith (1977). The cells were supplied with 5-nitrosalicylaldehyde (NSA) and amino acid derivatives of 4-methoxy-β-naphthylamine (MNA), which diffused readily into the cytoplasm. The liberated aromatic amine (MNA) (blue fluorescence at 425 nm) from peptidase activity formed an insoluble fluorescent Schiff-base complex (yellow–orange fluorescence with peaks at 530 and 595 nm) with NSA that was trapped inside the cells. Peptidase activity was estimated from the intensity of the fluorescence. This method has been tested on liquid cultures of the common soil ciliate *Colpoda steinii* with the bacterium *Pseudomonas fluorescens*. An EPICS CS flow cytometer (Coulter Instruments) was used to identify the ciliate population. Leucine–4-methoxy-β-naphthylamine (leucine MNA) and 5-nitrosalicylaldehyde (NSA) were applied at the concentrations used by Dolbeare and Smith (1977). After staining, the ciliates were fixed in glutaraldehyde (1% v/v). Some of the results are shown in Fig. 8.1 and Table 8.1. Unstained *C. steinii* cells fixed in 1% glutaraldehyde were readily distinguished from stained cells using a suitable gate and the green fluorescence detector. This stain of Dolbeare and Smith (1977) and the redox dye 5-cyano-2,3-ditolyl tetrazolium chloride (Rodriguez *et al.*, 1992) are two stains that could possibly be used to separate metabolically active soil protozoan populations by flow cytometry, providing a satisfactory technique for separating protozoans from soil particles can be discovered. More information is required about how protozoa and soil particles are attached to each other, if such a technique is to be developed quickly. Some of the existing methods for the selective dissolution of inorganic and organic components of soils, e.g. the use of sodium or potassium pyrophosphate (McKeague, 1967; Bascomb, 1968), acid ammonium oxalate (Schwertmann, 1973) and sodium dithionite (Mehra and Jackson, 1960; Woolson and Axley, 1969) may prove suitable for releasing protozoa from soil particles, when the cells are fixed beforehand and when very dilute preparations of the reagents are used. Cell biological techniques for gently separating small numbers of cells, e.g. settling velocity sediment at unit gravity or isokinetic sedimentation (Pretlow and Pretlow, 1982), electromigration (Borkott, 1975), immunocapture with either magnetic beads (Christensen *et al.*, 1992) or with non-cavitating ultrasound may also be required as part of preliminary purification procedures.

Soils often appear heterogeneous when examined at the microscopical scale and this may be reflected by some degree of aggregation in the distribution of soil protozoa. Geostatistical methods similar to those described by Webster and Boag (1992) for soil nematodes could be used to establish the degree of non-random distribution of soil protozoa. Another approach that could be used by soil protozoologists to reduce the difficulties posed

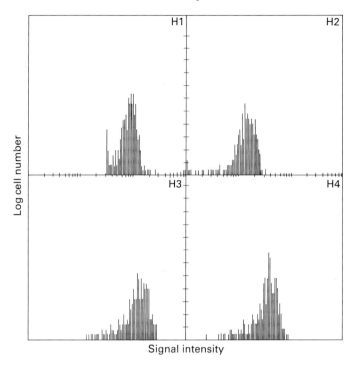

**Fig. 8.1.** Frequency distributions from four light detectors of an EPICS CS flow cytometer (Coulter Instruments) when used with a sample of the soil ciliate (*Colpoda steinii*) stained with a fluorescent Schiff-base complex from the intracellular reaction of 4-methoxy-β-naphthylamine (MNA) with 5-nitrosalicylaldehyde (NSA) (Dolbeare and Smith, 1977), and fixed with 1% (v/v) glutaraldehyde. Results from forward light scatter detector (FLS) are shown in histogram H1, red fluorescence detector (RFL) in histogram H2, green fluorescence detector (GFL) in histogram H3 and 90° light scatter detector (90LS) in histogram H4. All four signals were monitored on logarithmic scales. Further experimental details are given in legend to Table 8.1.

by soil heterogeneity is to consider what protozoan communities exist in localized regions of a particular soil. One such region or niche is the rhizosphere of living plants as discussed in Chapter 5. This niche could be extended for soil protozoa to include old root channels and associated root debris of dead plants or of harvested crops (Darbyshire *et al.*, 1977). Professor Hattori's suggestion in Chapter 3 to regard a soil aggregate, as a convenient unit for protozoan ecological studies, identified another obvious niche at a smaller scale. Other soil regions that may prove on examination to have distinct protozoan communities are the burrows of soil animals, zones immediately adjacent to organic matter deposits recently derived from plants, animals or microorganisms, the most superficial layer of soil and the

**Table 8.1.** Correlation coefficients between different signals from an EPICS CS flow cytometer (Coulter Instruments) when used to examine a sample of the soil ciliate (*Colpoda steinii*) stained with a fluorescent Schiff-base complex from the intracellular reaction of 4-methoxy-β-naphthylamine (MNA) with 5-nitrosalicylaldehyde (NSA) (Dolbeare and Smith, 1977), and fixed with 1% (v/v) glutaraldehyde. Forward light scatter signal (FLS), 90° light scatter signal (90LS), green fluorescence signal (515 to 530 nm, GFL) and red fluorescence signal (>590 nm, RFL). All four signals were monitored on logarithmic scales indicated by the 'L' prefix in front of the detector abbreviations. The argon ion laser was operated at 458 nm wavelength with a power output of 100 mW. Low level interference from bacteria was removed from the LFLS signal by using a suitable gate. A neutral density filter was inserted in front of the FLS detector to reduce signal strength. Sensitivities for 90LS, RFL and GFL detectors were set at 150v, 500v and 450v, respectively.

|        | L90LS | LGFL | LRFL |
|--------|-------|------|------|
| LFLS   | 0.67  | 0.67 | 0.74 |
| L90LS  |       | 0.95 | 0.80 |
| LGFL   |       |      | 0.78 |

capillary versus non-capillary soil pores. The linings of earthworm burrows are known to have distinct chemical and fungal characteristics (Jeanson, 1972) and there is some evidence that earthworm activities in soil stimulate the growth of microbial populations (Scheu, 1987; Daniel and Anderson, 1992). The superficial layers of many soils when unprotected by litter layers are harsh environments and are liable to crust formation or erosion (Slattery and Bryan, 1992). Protozoa in these layers need to be resistant to rapid environmental changes and it is possible that few active trophozoites exist in this region. According to a recent simulation model of solute leaching (Barraclough, 1989), two processes apart from 'normal' connective–dispersive flow were involved: preferential flow in which solute moving in the larger water-filled pores does not equilibrate with that in the smaller pores (Van Genuchten and Wierenga, 1976) and by-pass flow in which water and solute move rapidly down gravitationally drained cracks and macropores without displacing water in the soil matrix (Thomas and Phillips, 1979). Apart from its importance for the movement of fertilizers, pesticides or other toxic chemicals through soil, such water flow may have effects on the distribution of protozoa that have not been investigated. Protozoa in the smaller pores may often be surrounded by nearly stagnant soil solution (Van Genuchten and Wierenga, 1976; Skopp *et al.*, 1981). Future investigations may demonstrate differences between protozoan populations in smaller and larger pores that reflect these environmental conditions.

At an even smaller scale, one can only conjecture at present about the effects of changes in soil surfaces and soil solutions on protozoan colonization and feeding. Artificial soil aggregates composed of different surfaces and pore sizes may help to distinguish between surface and pore effects (Darbyshire *et al.*, 1993). Roper and Marshall (1979) stated, in a study of a Tasmanian river, that there were two possible interactions between microorganisms and particulate matter in that river: (i) sorption of microorganisms to large surfaces or aggregates; (ii) sorption of colloidal particulates to the surfaces of microbes. In their freshwater system, much of the particulate matter was present as suspended colloids and so capable of attachment to microbial surfaces by electrostatic attraction. As salinity increased in the lower reaches of the estuary, flocculation and sedimentation of the microorganisms and particles occurred. Evidence that similar phenomena occur in soils is provided by the appearance of microorganisms in soil in electron micrographs (Kilbertus, 1980; Foster, 1988; Robert and Chenu, 1992) and from reports that the ionic strength of soil solutions has a significant effect on the transport of bacteria through soil columns (Tan *et al.*, 1991, 1992). In these latter studies, it was inferred that the ionic strength of the soil solution affected the adhesion of bacteria to soil particles. It is also possible that changes in the strength of adhesion of microorganisms to soil surfaces may influence the ability of protozoa to eat such prey and recycle the nutrients.

Existing autecological information about soil protozoa is meagre. Pratt (1992) distinguished three protozoan feeding groups in aquatic environments: filter feeders (sedentary or planktonic), surface gliders that pick up individual particles from surfaces and protozoa that actively intercept prey. Filming the behaviour of soil representatives from these three feeding groups in soil microcosms may provide new insights into protozoan feeding and colonization, especially if the model systems are realistic representations of soil conditions with aggregates and microorganisms isolated from soil. The use of epifluorescence microscopy and low-light-sensitive videocameras should make it possible to use near-natural densities of soil aggregates in these laboratory systems. Fluorescent or luminescent gene probes could be used to follow changes in protozoan populations. Comparisons between the behaviour of protozoan mutants deficient in some characteristic and wild strains may also be useful in such studies.

Obviously, many intricacies of soil protozoan ecology remain to be investigated and this task represents an exciting challenge to protozoologists. The recent improvements in knowledge and techniques discussed in this book extend the range of possible research topics. If this challenge is accepted by even a small number of protozoologists, preferably associated with multidisciplinary groups of microbiologists and soil scientists, then significant advances should be made in the next decade. The benefits would extend well beyond soil protozoology. They would improve our appreciation of the importance of soil organisms in the maintenance of soil fertility and therefore indirectly in human welfare.

# Acknowledgements

The flow cytometry results shown in Fig. 8.1 and Table 8.1 were provided by Professor T.R.G. Gray, Dr M. Marshall and Phillip Reed of the Department of Biology, University of Essex. I am very grateful for their help.

# References

Amman, R.I., Binder, B.J., Olson, R.J., Chisholm, S.W., Devereux, R. and Stahl, D.A. (1990) Combination of 16S rRNA-targeted oligonucleotide probes with flow cytometry for analysing mixed microbial populations. *Applied and Environmental Microbiology* 56, 1919–1925.

Barraclough, D. (1989) A usable mechanistic model of nitrate leaching I. The model. *Journal of Soil Science* 40, 543–554.

Bascomb, C.L. (1968) Distribution of pyrophosphate-extractable iron and organic carbon in soils of various groups. *Journal of Soil Science* 19, 251–268.

Borkott, H. (1975) A method for quantitative isolation and preparation of particle-free suspensions of bacteriophagous ciliates from different substrates for electronic counting. *Archiv für Protistenkunde* 117, 261–268.

Burkill, P. (1992) Analytical flow cytometry in marine protozoan research. *NERC News* 20, 5–7.

Christensen, B., Torsvik, T. and Lien, T. (1992) Immunomagnetically captured thermophilic sulfate-reducing bacteria from North Sea oil field waters. *Applied and Environmental Microbiology* 58, 1244–1248.

Corliss, J.O. (1992) The interface between taxonomy and ecology in modern studies on the protists. *Acta Protozoologica* 31, 1–9.

Daniel, O. and Anderson, J.M. (1992) Microbial biomass and activity on contrasting soil materials after passage through the gut of the earthworm *Lumbricus rubellus* Hoffmeister. *Soil Biology and Biochemistry* 24, 465–470.

Darbyshire, J.F., Davidson, M.S., Scott, N.M. and Shipton, P.J. (1977) Some microbial and chemical changes in soil near the roots of spring barley, *Hordeum vulgare* L., infected with take-all disease. *Ecological Bulletins (Stockholm)* 25, 374–380.

Darbyshire, J.F., Chapman, S.J., Cheshire, M.V., Gauld, J.H., McHardy, W.J., Paterson, E. and Vaughan, D.V. (1993) Methods for the study of interrelationships between microorganisms and soil structure. *Geoderma* 56, 3–23.

Dolbeare, F.A. and Smith, R.E. (1977) Flow cytometric measurement of peptidases with use of 5-nitrosalicylaldehyde and 4-methoxy-β-napthylamine derivatives. *Clinical Chemistry* 23, 1485–1491.

Edwards, C., Porter, J., Saunders, J.R., Diaper, J., Morgan, J.A.W. and Pickup, R.W. (1992) Flow cytometry and microbiology. *Society for General Microbiology Quarterly* 19, 105–108.

Foster, R.C. (1988) Microenvironments of soil microorganisms. *Biology and Fertility of Soils* 6, 189–203.

Jeanson, C. (1972) Étude microscopique de depôts de fer, de manganèse et de calcium dans un sol experimental – leur association avec microorganisms. *Revue d'Écologie et de Biologie du Sol* 9, 479–489.

Kilbertus, G. (1980) Étude des microhabitats contenus dans les agrégats du sol, leur relation avec la biomasse bacterienne et la taille des procaryotes presents. *Revue d'Écologie et de Biologie du Sol* 17, 543–557.

McKeague, J.A. (1967) An evaluation of 0.1 M pyrophosphate and pyrophosphate-dithionite in comparison with oxalate as extractants of the accumulation products in podzols and some other soils. *Canadian Journal of Soil Science* 47, 95–99.

Mehra, O.P. and Jackson, M.L. (1960) Iron oxide removal from soils and clays by a dithionite-citrate system buffered with sodium bicarbonate. *Proceedings of 7th National Conference on Clays and Clay Minerals* pp. 317–327.

Pratt, J.R. (1992) Protozoan feeding groups revisited: adaptations to surface feeding. *Journal of Protozoology* 39, Supplement 10A (abstract 58).

Pretlow, T.G. and Pretlow, T.P. (1982) Sedimentation of cells: an overview and discussion of artifacts. In: Pretlow, T.G. and Pretlow, T.P. (eds), *Cell Separation Methods and Selected Applications*, Vol. 1. Academic Press, London, pp. 41–60.

Robert, M. and Chenu, C. (1992) Interactions between soil minerals and microorganisms. In: Stotzky, G. and Bollag, J. (eds), *Soil Biochemistry*, Vol. 7. Marcel Dekker, New York, pp. 307–404.

Rodriguez, G.G., Phipps, D., Ishiguro, K. and Ridgway, H.F. (1992) Use of a fluorescent redox probe for direct visualization of actively respiring bacteria. *Applied and Environmental Microbiology* 58, 1801–1808.

Roper, M.M. and Marshall, K.C. (1979) Effects of salinity on sedimentation and of particulates on survival of bacteria in estuarine habitats. *Geomicrobiology Journal* 1, 103–116.

Scheu, S. (1987) Microbial activity and nutrient dynamics in earthworm casts (Lumbricidae). *Biology and Fertility of Soils* 5, 230–234.

Schwertmann, U. (1973) Use of oxalate for Fe extraction from soils. *Canadian Journal of Soil Science* 53, 244–246.

Shapiro, H.M. (1990) Flow cytometry in laboratory microbiology: new directions. *American Society of Microbiology News* 56, 584–588.

Skopp, J., Gardner, W.R. and Tyler, E.J. (1981) Solute movement in saturated soils: Two-region model with small interaction. *Soil Science Society of America Journal* 45, 837–842.

Slattery, M.C. and Bryan, R.B. (1992) Laboratory experiments on surface seal development and its effect on interrill erosion processes. *Journal of Soil Science* 43, 517–529.

Tan, Y., Bond, W.J., Rovira, A.D., Brisbane, P.G. and Griffin, D.M. (1991) Movement through soil of a biological control agent, *Pseudomonas fluorescens*. *Soil Biology and Biochemistry* 23, 821–825.

Tan, Y., Bond, W.J. and Griffin, D.M. (1992) Transport of bacteria during unsteady unsaturated soil water flow. *Soil Science Society of America Journal* 56, 1331–1340.

Thomas, G.W. and Phillips, R.E. (1979) Consequence of water movement in macropores. *Journal of Environmental Quality* 8, 149–152.

Van Genuchten, M.T. and Wierenga, P.J. (1976) Mass transfer studies in porous media. I. Analytical solution. *Soil Science Society of America Journal* 40, 473–480.

Webster, R. and Boag, B. (1992) Geostatistical analysis of cyst nematodes in soil. *Journal of Soil Science* 43, 583–595.

Woolson, E.A. and Axley, J.H. (1969) Clay separation and identification by a density gradient procedure. *Soil Science Society of America Proceedings* 33, 46–48.

# Index